Studies in Computational Intelligence

Volume 1217

The series "Studies in Computational Intelligence" (SCI) publishes new developments and advances in the various areas of computational intelligence—quickly and with a high quality. The intent is to cover the theory, applications, and design methods of computational intelligence, as embedded in the fields of engineering, computer science, physics and life sciences, as well as the methodologies behind them. The series contains monographs, lecture notes and edited volumes in computational intelligence spanning the areas of neural networks, connectionist systems, genetic algorithms, evolutionary computation, artificial intelligence, cellular automata, self-organizing systems, soft computing, fuzzy systems, and hybrid intelligent systems. Of particular value to both the contributors and the readership are the short publication timeframe and the world-wide distribution, which enable both wide and rapid dissemination of research output.

Indexed by SCOPUS, DBLP, WTI AG (Switzerland), zbMATH, SCImago.

All books published in the series are submitted for consideration in Web of Science.

Rodrick Wallace

Hallucination and Panic in Autonomous Systems

Paradigms and Applications

Springer

Rodrick Wallace
Division of Epidemiology
The New York State Psychiatric Institute
Columbia University
New York, NY, USA

ISSN 1860-949X ISSN 1860-9503 (electronic)
Studies in Computational Intelligence
ISBN 978-3-031-95206-7 ISBN 978-3-031-95207-4 (eBook)
https://doi.org/10.1007/978-3-031-95207-4

This Springer imprint is published by the registered company Springer Nature Switzerland AG
The registered company address is: Gewerbestrasse 11, 6330 Cham, Switzerland

Preface

> *Alchemists were so successful in distilling quicksilver from what seemed to be dirt, that after several hundred years of fruitless effort to convert lead into gold they still refused to believe that on the chemical level one cannot transmute metals. To avoid the fate of the alchemists, it is time we asked where we stand.*
>
> —H. Dreyfus (1965)

> *...So, are they saying Pinocchio becomes a real boy?*
>
> —D. N. Wallace (2024)

Most current attempts to build 'intelligent machines', ranging from relatively simple Large Language Model chatbots to an 'Artificial General Intelligence' that would rival or surpasses human minds, are blindsided by the reality that all such machines are, and always will be, cultural artifacts of those who build them, reflecting embedded cultural traits, patterns, and historical trajectories. In particular, the Western 'Tech Bro' mindset that dominates current research inevitably focuses on salient objects at the expense of context (e.g., Nisbett et al. 2001). For intelligence-as-we-know-it, however, context is what things are really all about.

That is, intelligence arose under the evolutionary selection pressure of rapid changes in embedding environmental patterns of threat and affordance (Gould 2002; Wallace 2022, 2023). The medium for response to those embedding dynamics is, of course, embodiment. No hyperrapid embodiment, involving close feedback coupling with a rich embedding world, no real 'intelligence' (Fjelland 2020; Dreyfus 1972, 1986). Simple, direct, compelling, and diametrically opposite to present bizarrely narrow Western Tech Bro venture capital thinking.

As Iida and Giardina (2023) put it,

> Embodiment is a crucial concept for the autonomy and adaptivity of systems working in the physical world with high degrees of uncertainty and complexity. The physical bodies of autonomous adaptive systems heavily influence the information flow from the environment to the central processing (and vice versa), requiring us to consider the full triad of brain, body, and environment to investigate intelligent behavior.

Barrett and Stout (2024), as part of an extensive review, further comment that

> …[T]here is an interesting convergence across ‘embodied’ perspectives on cognition, culture and evolution, all of which seek to move away from simplifying software/hardware, mind/body, organism/environment and cause/mechanism dichotomies of the mid-twentieth-century information-processing paradigm. The result is a more integrated approach to cognition as emerging from dynamic interactions across behavioural, developmental, historical and evolutionary timescales…
>
> …Progress will require the development of new methods for the rich recording and analysis of human behaviour in real-world contexts combined with a deeper appreciation of the complex interactions between bio-behavioural mechanisms, social dynamics and cultural systems of meaning. Relevant processes range in scale from neurons to societies and call for an approach that is both inclusive and integrative, embracing quantitative, qualitative and applied research, focused experiments as well as long-term observations and detailed case studies as well as broad comparative work.

Here, we address these questions—most particularly the ‘development of new methods’—in terms of the asymptotic limit theorems of information and control theories that, in the sense of Dretske (1994), necessarily constrain all cognitive phenomena. This is a deliberately limited perspective, aimed at providing both scientific insight and, most specifically, new statistical tools for the study of observational and experimental data regarding cognition, embodiment, intelligence, and their many dysfunctions and pathologies.

As these have proven to be treacherous grounds—for some, literally invoking Dreams of Reason—an extended commentary is warranted.

Regarding the many—and growing—misperceptions, misuses, and mysticisms of ‘information’, Longo (2020) writes

> The modern theories of information opened the way to extraordinary new tools for communication and knowledge construction. When identified instead with the world (DNA is a complete information carrier and a program for ontogenesis, atomic particles and planets exchange information…), a new form of scientism affects knowledge and trust in science. Vague and implicit assumptions, with no precise references to the sciences of information, force strong consequences, in biology in particular…
>
> As it often happens, the mystics have taken the beautiful mathematical invariance in the… [Turing and Shannon] theories of information as an absolute and have applied it to all phenomena, including natural ones…

Rigorous information and control theories—as we attempt to address them here—are encompassed by a remarkably small set of severely constraining asymptotic limit theorems and associated lemmas (Khinchin 1957; Cover and Thomas 2006; Nair et al. 2007):

- The Shannon Coding Theorem that prescribes how to code a message in ‘typical’ sequences for maximum error-free transmission through a particular channel at a maximum ‘channel capacity’ (and the ‘tuning theorem’ variant in which the message formally ‘transmits’ the channel).
- The Source Coding Theorem that splits a set of sequences emitted by an information source into two subsets, a relatively small one that is ‘meaningful’—the sequences are consonant with an underlying ‘grammar’ and ‘syntax’—and a much

larger 'meaningless' one 'of measure zero'. Only for stationary, ergodic systems is it possible to characterize 'information' and 'information source uncertainty' in terms of a Shannon entropy analog. For nonergodic systems, each (infinite) signal sequence has it's own scalar source uncertainty that cannot be expressed in 'Shannon entropy' form (Khinchin 1957). It is here that methods must be abducted from statistical physics, introducing as well equivalence classes of sequences forming a groupoid, leading to groupoid symmetry-breaking phase transitions unlike those of 'conventional' physical systems.

- The Rate Distortion Theorem (RDT) and an associated Rate Distortion Control Theory that, in a sense, 'tunes' the channel capacity around a maximum-permissible scalar measure of average distortion (which finesses the Friston 'free energy minimization' approach).
- The Data Rate Theorem (DRT) stating that, for stability of an inherently unstable control system, 'control information' must be provided at a rate greater than the system produces it's own 'topological information'. A version of this, we will show, can be 'easily' derived from a time and 'free energy rate' optimization based on the Hartley-Shannon Theorem.

After some further development:

- Cognition implies choice—of a small set of actions from a larger embedding set of what is possible to the system of interest. Such choice reduces uncertainty in a formal manner, implying the existence of an information source 'dual' to the cognitive process studied, at and across the varied scales and levels of organization of the living state (Atlan and Cohen 1998; Dretske 1994; Maturana and Varela 1980).
- Real-world, real-time cognition is always inherently unstable under the impact of the 'topological information' imposed by embedding environmental 'roadways'. Invocation of the DRT then implies that *cognition must always be paired with regulation for system stability and long-term function under selection pressures.*
- We implicitly assume an adiabatically, piecewise, stationary (APS) 'Born-Oppenheimer' model in which the system of interest remains 'close enough' to stationary during the period(s) of interest so that the asymptotic limit theorems are effectively 'anytime algorithms' (e.g., Zilberstein 1996). Between 'pieces' for which this assumption is sufficiently true, we invoke (groupoid) symmetry-breaking and/or Fisher Zero (Dolan et al. 2001) phase transitions.
- The ergodic restriction, we will show, can 'easily' be lifted, at the cost of imposing methods abducted from statistical physics, with dynamics approximated via an analog to the first (and higher) order Onsager treatment of nonequilibrium thermodynamics, recognizing, however, that, due to the failure of information source microreversibility—e.g., 'eht' does not have the same probability as 'the' in English—there are no 'Onsager reciprocal relations' in high order approximations.
- For purely physical phenomena, free energy is often defined in terms of Boltzmann distribution-derived partition functions. In contrast, free energy analogs associated

with the dynamics of cognitive phenomena may well be characterized by 'fat-tailed' distributions that do not have well-defined first or second moments (e.g., Derman et al. 2016; Wallace and Fricchione 2024).

A clever postdoc could take these points and, over a long weekend, derive much of the interesting theory that follows. However, moving much beyond this formal landscape leads to realms of mysticism—like Tononi's (and Tegmark's) infamous 'Integrated Information Theory' (Fleming et al. 2023)—that are, in physics terminology, 'Not Even Wrong'.

A good deal of current discourse on 'information', cognition, consciousness, and biology, sadly to say, lies in the direction of the comments by H. Dreyfus and D. N. Wallace that open this Preface.

It is unlikely that 'LLMs', 'driverless cars', 'AGI', 'autonomous weapons', 'AI General Officers', and the like, can escape the limitations imposed by the ultimate necessities of embodiment-and-regulation for stable and 'trustworthy' cognition.

As Acemoglu (2024) put it,

> Many tasks that humans currently perform, for example in the areas of transportation, manufacturing, mining, etc., are multifaceted and require real-world interaction, which AI won't be able to materially improve anytime soon…

Watts (2008) places some of this in a military context:

> Tactical problems are 'tame' in that they generally have definite solutions in an engineering sense. So-called 'wicked' problems are fundamentally social ones. They are ill-structured, open-ended, and not amenable to closed, engineering solutions. Operational and strategic problems appear to lie within the realm of wicked or messy problems. Such problems give every indication of requiring hard thinking and careful analysis rather than quick, intuitive responses.

Wallace (2021) continues the line of argument:

> The physical and the cognitive mesh and clash in a dynamic syncretism that includes, not only the overwhelming influences of culture and path-dependent historical trajectory, but the inevitable burdens of Clausewitzian fog and friction. Such factors sculpt conflict at all scales and levels of organization. …[I]nstitutional cognition—always and inevitably embodied—even when 'enhanced' by artificial intelligence or related machineries, must suffer significantly increased probability of failure under conditions of fog and friction, particularly for 'wickedly hard' tactical, operational, and strategic problems not amenable to straightforward, if difficult, engineering solutions.

While the US-based corporation Palantir provides an AI system that appears to contribute significantly to tactical efforts by the Ukraine in the current war with Russia, strategic matters in that war remain grindingly—and traditionally—wicked (e.g., Rickli and Mantellasi 2024).

Indeed, cultural aspects of strategic wickedness are a recurring theme among US military planners. As Johnson (2022) put it,

> …U.S. defense institutions have repeated a pattern of investing in cultural study during short bursts of counterinsurgency fighting and then abandoning it along with its lessons learned at the termination of conflict. As a consequence, U.S. planning efforts—including those now

> being designed for future great power conflict—suffer from an unnecessarily narrow optic and fail to account for the full range of perspectives and plausible courses of action considered by an adversary. America's allies know it and are frustrated by it. More importantly, U.S. adversaries know it and plan to exploit it.

To reiterate, the focus here is on probability models derived from the fundamental asymptotic limit theorems of information and control theories, extended by standard methods abducted from statistical physics and nonequilibrium thermodynamics as needed. In addition to exploring scientific insights such models might bring, a specific aim is the development of new statistical tools for the analysis of observational and experimental data regarding embodied cognition and, in particular, the dynamics of its dysfunctions and their possible remediation. Such data are, after all, the only sources of new knowledge, as opposed to new speculation.

Chapter 1 adapts and applies statistical mechanics and nonequilibrium thermodynamics formalisms to Rate Distortion Control Theory, deriving the Yerkes-Dodson inverted-U signal transduction as a function of 'arousal', finding that 'hallucination' modes are particularly driven by a disjunction between system time constant and the 'natural' time constants of embedding selection pressures.

Chapter 2 reprises the latter result, using a simplified methodology focusing on the Hartley-Shannon theorem. In essence, one recovers John Boyd's OODA loop dynamic as a disjunction between the characteristic time of the embodied system under study and that of the embedding environment.

Chapter 3 examines the intertwined dynamics of cognition and regulation at some length, expanding recent results from Wallace (2025, Ch. 3) on 'innate' and 'acquired' regulation.

Chapter 4, adapted from Wallace (2022), reconsiders the formalism, taking 'culture' itself as part of a larger, joint information source, eventually recovering the earlier development as involving a distortion measure defined by the rate of control information needed to stabilize an inherently unstable system.

Chapter 5, expanding on Wallace (2021), explores the dynamics of 'wicked' problems faced by inherently embodied institutions engaged in organized conflict.

Chapter 6 applies the model to the canonical instabilities of AI-enabled High Frequency Trading.

Chapter 7 explores extension of the model when essential resources cannot be projected down to a single scalar composite, and related matters.

Chapter 8 models the effects of often critically necessary data compression schemes on cognitive dynamics, showing how Weber-Fechner and Stevens Law condensations drive the Yerkes-Dodson Effect, and how sensitivity and compression can interact to trigger gross instabilities.

Chapter 9 explores the difference between Detailed Command and Mission Command, using a cancer-stage model.

Chapter 10 places the analysis in context: human minds, institutions, and machine systems are all cultural artifacts.

The chapters, laced with worked-out examples, can be read independently, at the expense of some repetition.

A skilled reader can take this material and run with it.

New York, USA Rodrick Wallace

References

Acemoglu, D., 2024, Interview with Daron Acemoglu, Top of Mind, Issue 129, Goldman-Sachs Global Investment Research.

Atlan, H., I. Cohen, 1998, Immune information, self-organization, and meaning, International Immunology, 10:711–717.

Barrett, L., D. Stout, 2024, Minds in movement: embodied cognition in the age of artificial intelligence, Philosophical Transactions B 379:20230144, https://doi.org/10.1098/rstb.2023.0144.

Cover, T., J. Thomas, 2006, Elements of Information Theory, Second Edition, Wiley, New York.

Derman, E., M. Miller, D. Park, 2016, The Volatility Smile, Wiley, New York.

Dolan, B., W. Janke, D. Johnston, M. Stathakopoulos, 2001, Thin Fisher zeros, Journal of Physics A 34:6211–6223.

Dretske, F., 1994, The explanatory role of information, Philosophical Transactions of the Royal Society A 349:59–70.

Dreyfus, H., 1965, Alchemy and Artificial Intelligence, Rand Report P-3244.

Dreyfus, H., 1972, What computers can't do. Harper and Row, New York.

Dreyfus, H., S. Dreyfus, 1986, Mind Over Machine, Blackwell, Basil.

Fjelland, R., 2020, Why general artificial intelligence will not be realized, Humanities and Social Sciences Communications, 7:10, https://doi.org/10.1057/s41599-020-0494-4.

Fleming et al., 2023, The integrated information theory of consciousness is pseudoscience, https://osf.io/preprints/psyarxiv/zsr78.

Gould, S.J., 2002, The Structure of Evolutionary Theory, Harvard University Press, Cambridge MA.

Iida, F., F. Giardina, 2023, On the timescales of embodied intelligence for autonomous adaptive systems, Annual Review of Control, Robotics, and Autonomous Systems 6:95–122.

Johnson, J., 2022, From the Editor, Journal of Advanced Military Studies Special Issue on Strategic Culture, https://doi.org/10.21140/mcuj.2022SIstratcul001.

Khinchin, A., 1957, The Mathematical Foundations of Information Theory, Dover, New York.

Longo, G., 2020, Information, science and democracy: For an ethics of scientific knowledge, In A. Fabris (ed.), Trust, Studies in Applied Philosophy, Epistemology and Rational Ethics 54, Springer, New York.

Maturana, H., F. Varela, 1980, Autopoiesis and Cognition: The Realization of the Living, Reidel, Boston.

Nair, G., F. Fagnani, S. Zampieri, R. Evans, 2007, Feedback control under data rate constraints: an overview, Proceedings of the IEEE 95, 108138.

Nisbett, R., K. Peng, C. Incheol, A. Norenzayan, 2001, Culture and systems of thought: holistic vs. analytic cognition, Psychological Review 108:291–310.

Rickli, J., F. Mantellasi, 2024, The war in Ukraine: Reality check for emerging technologies and the future of warfare, Geneva Paper 34/24, Geneva Centre for Security Policy.

Wallace, D.N., 2024. Personal communication.

Wallace, R., 2021, Embodied cognition and its pathologies: The dynamics of institutional failure on wickedly hard problems, Communications in Nonlinear Science and Numerical Simulation, 95:105616.

Wallace, R., 2022, Consciousness, Cognition and Crosstalk: The evolutionary exaptation of nonergodic groupoid symmetry-breaking, Springer, New York.
Wallace, R., 2023, Essays on the Extended Evolutionary Synthesis: Formalizations and expansions, Springer, New York.
Wallace, R., 2025, Mathematical Essays on Embodied Cognition: Insights from Information and Control Theories, Springer, New York.
Wallace, R., G. Fricchione, 2024, Stress-induced failure of embodied cognition: a general model, BioSystems, https://www.sciencedirect.com/science/article/abs/pii/S0303264724000789.
Watts, B., 2008, US Combat training, operational art, and strategic competence: Problems and opportunities. Washington, DC: Center for Strategic and Budgetary Assessments.
Zilberstein, S., 1996, Using anytime algorithms in intelligent systems, AI Magazine, 17:73–83.

Contents

About the Author

Rodrick Wallace is a research scientist in the Division of Epidemiology at the New York State Psychiatric Institute, affiliated with the Columbia University Department of Psychiatry. He has an undergraduate degree in mathematics and a Ph.D. in physics from Columbia and completed postdoctoral training in the epidemiology of mental disorders at Rutgers. He worked as a public interest lobbyist, including two decades conducting empirical studies of fire service deployment, and subsequently received an Investigator Award in Health Policy Research from the Robert Wood Johnson Foundation. In addition to material on public health and public policy, he has published peer-reviewed studies modeling evolutionary process and heterodox economics, as well as many quantitative analyses of institutional, machine, and biological cognition. He publishes extensively in the military science literature and received one of the UK MoD RUSI Trench Gascoigne Essay Awards.

Chapter 1
'Machine Intelligence' Is an Oxymoron Without Real-Time Embodiment and Regulation

Abstract Elementary formal models based on the asymptotic limit theorems of information and control theories challenge current culturally and ideologically-constrained Western 'Tech Bro' conceptions of Artificial General Intelligence. Confronted by Darwinian selection, cognition has always been paired with real-time regulation under an embodiment that interacts with, but is part of, an often actively adversarial embedding environment. Absent real-time embodiment/regulation, there can be no intelligence as-we-know-it. In particular, Western Tech Bro atomistic cognitive machines will not function well confronting the 'Clausewitz landscapes' of fog, friction, and adversarial intent that characterize much of real-world experience. As history has shown, Darwinian selection challenges the culture-bound ideologies of even the most powerful empires. The modeling approach, which is surprisingly general, can be used to construct specialized statistical tools for the analysis of observational and experimental data regarding cognition across a broad range of scales and levels of organization.

> ***Recent unexpected difficulties in realising truly autonomous vehicles, making reliable medical diagnoses, detecting offensive online content and even just recognising faces, show that brains remain significantly functionally more capable than we can currently emulate. …[W]e must embrace the brain's inherent dynamic complexity and emergent properties and explain how plasticity moulds the dynamics to capture useful couplings across brain regions and between the brain, the body, and the world.***
>
> —Stratton (2024).

> ***…[M]achine learning appears to have run into many of the same problems of brittleness and lack of versatility as more traditional engineering approaches when deployed on a robot operating outside well-defined conditions such as the warehouse, factory floor, or carefully mapped streets in a city with an unvarying, sunny climate.***
>
> —Roy et al. (2021).

R. Wallace, *Hallucination and Panic in Autonomous Systems*, Studies in Computational Intelligence 1217, https://doi.org/10.1007/978-3-031-95207-4_1

1.1 Introduction

Our machines, of whatever nature and form, are cultural artifacts that, like woven baskets, reflect—and inhabit—our embedding sociocultural world. Current Silicon Valley Tech Bro views of, and attempts at, 'Artificial General Intelligence' (AGI), are, and will be, constrained by the cultural perspective of those building it. Indeed, persistent Tech Bro conceptions of AGI have long been seen as pathological Dreams of Reason (e.g., Dreyfus, H., 1972, 1972; Fjelland 2020).

The current, and ongoing, driverless car fiasco provides a window into the consequences of constrained cultural perspectives. Wallace (2007) writes,

> The asymptotic limit theorems of control and information theories make it possible to explore the dynamics of collapse likely to afflict systems of autonomous ground vehicles that communicate with each other and with an embedding intelligent roadway. A vehicle/road system is inherently unstable in the control theory sense as a consequence of the basic irregularities of the traffic stream, the road network, and their interactions, placing it in the realm of the data rate theorem that mandates a minimum necessary rate of control information for stability. It appears that such V2V/V2I systems will experience large-scale failures analogous to the vast propagating fronts of power network blackouts, and possibly less benign, but more subtle patterns of individual, platoon, and mesoscale dysfunction.
>
> An atomistic perspective on autonomous ground vehicles – seeing them as having only local dynamics in an embedding traffic stream – embodies a profound failure of insight. Traffic light strategies, road quality, the inevitably rapid-shifting 'road-map space', the dynamic composition of the traffic stream, communication and machine sensory system bandwidth limits, and so on, create the synergistic context in which single vehicles operate. It is necessary to understand the dynamics of that full system, not simply the behavior of an individual vehicle atom within it. The properties of that system will be both overtly and subtly emergent – subject to sudden 'phase transitions' into both massively and locally unstable modes – as will the responses of individual cognitive vehicles enmeshed in context, whether controlled by humans or computers. The triggering of adverse events at various scales and levels of organization by unfriendly external agents will likely become routine....

We will assert, in parallel with this analysis, that attempts to construct intelligent systems under Tech Bro cultural groupthink, with its extreme—often pathological—isolation and siloing of salient signals and events at the expense of context, are bound to follow a similar path-dependent trajectory. Corrective measures—a formal alternative strategic perspective—will be outlined below.

1.2 The Tech Bro Cultural Rabbithole

A broad range of human studies regards the role of culture in determining brain hard-wiring for basic perception (Wallace and Wallace 2016). In particular, Nisbett et al. (2006), consonant with a long line of research (e.g., Markus and Kitayama 1991; Heine 2001), review an extensive range of empirical studies regarding basic cognitive differences between individuals raised in East Asian and Western cultural heritages. These are characterized, respectively, as 'holistic' and 'analytic'. Their argument is roughly as follows:

- Social organization directs attention to some aspects of the perceptual field at the expense of others.
- What is attended to influences metaphysics.
- Metaphysics guides tacit epistemology, that is, beliefs about the nature of the world and causality.
- Epistemology dictates the development and application of some cognitive processes at the expense of others.
- Social organization can directly affect the plausibility of metaphysical assumptions, such as whether causality should be regarded as residing in the field versus in the object.
- Social organization and social practice can directly influence the development and use of cognitive processes such as dialectical versus logical ones.
- Thus tools of thought embody a culture's intellectual history, tools have theories built into them, and users accept these theories, often unknowingly, when they use these tools.

Masuda and Nisbett (2006), somewhat later, argue that research on perception and cognition suggests that whereas East Asians view the world holistically, attending to the entire field and relations among objects, Westerners view the world analytically, focusing on the attributes of salient objects. Compared to Americans, East Asians were more sensitive to contextual changes than to focal object changes, These results suggest that there can be cultural variation in what may seem to be basic perceptual processes. Nisbett and Miyamoto (2005) argue that fundamental perceptual processes are influenced by culture.

These findings establish a dynamic relationship between the cultural context and perceptual processes, implying strongly that perception can no longer be regarded as consisting of processes that are universal across all people at all times.

Wallace (2007), somewhat in parallel, explores analogous dynamics involving inattentional blindness and culture (Simons and Chabris 1999).

Broadly, then, within human populations there can be no single model of cognition and consciousness. This suggests, in turn, that those seeking to build machines that mimic or extend human cognitive abilities should be prepared to explore a very wide dynamic landscape using a very adaptable toolbox. In particular, the idea of an 'AGI' residing within some machine, embedded in a data stream that it interprets and commands, is a very Western cultural trope that may have very limited actual use.

In sum, a Tech Bro atomistic 'AGI'—inherently seen as, and thus built to be, separate from its embedding milieu—would be as fragile and useless, and as much a La Brea money pit, as the current generation of driverless cars and madly hallucinating Large Language Models.

1.3 Embodiment

The first step is to recognize that cognitive systems—biological, institutional, mechanical, or composite—are embedded in an environment that includes themselves. There will be (at least) three resource streams necessary for successful function. The first is the rate at which elements of the system can communicate with each other, instantiated by some channel capacity $\mathscr{C}$. The second is the rate at which 'sensory information' is available from the embedding circumstance, say according to a channel capacity $\mathscr{H}$. Finally, there is the rate at which 'material/materiel' resources can be provided, a rate $\mathscr{M}$. For an organism, this might be measured by the rate at which metabolic free energy is provided. Machine/institutional systems will have a different measure.

In general, these three rates will 'crosscorrelate' in a sense, so that one is actually confronting a 3 by 3 (or greater) matrix $\mathbf{Z}$. An $n \times n$ matrix will have will have n scalar invariants determined by the polynomial relation

$$\begin{aligned} p(\gamma) &= \det[\mathbf{Z} - \gamma \mathbf{I}] \\ &= (-1)^n \gamma^n + (-1)^{n-1} r_1 \gamma^{n-1} + \cdots - r_{n-1}\gamma + r_n \end{aligned} \tag{1.1}$$

$\mathbf{I}$ is the $n \times n$ identity matrix, det the determinant, and γ a real-valued parameter. The first invariant is the matrix trace, and the last the determinant.

The most direct index might be $Z = \mathscr{C} \times \mathscr{H} \times \mathscr{M}$, but the real world is not likely to be so simple. In any event, we now postulate a scalar index Z representing information and material resource rates. Wallace (2021) outlines generalization to more than one such index.

A more fully embodied formalism emerges via Rate Distortion Control Theory based on the Rate Distortion Theorem (RDT) (Cover and Thomas 2006). The RDT addresses the question of message transmission under conditions of noise for an information channel. There will be, for that channel, given a particular scalar measure of average distortion, D, between what is sent and what is received, a minimum necessary channel capacity $R(D)$.

Effectively, the theorem asks what is the 'best' channel for transmission of a message with the least possible average distortion. $R(D)$ can be defined for nonergodic information sources via a limit argument based on the ergodic decomposition of a nonergodic source into a 'sum' of ergodic sources. The RDT can be reconfigured in a control theory context, if we envision a system's topological information from the Data Rate Theorem as another form of noise, adding to the average distortion D. See Fig. 1.1. The punctuation implied by the less well-known Data Rate Theorem of control theory (Nair et al. 2007) emerges if there is a critical maximum average distortion beyond which control fails.

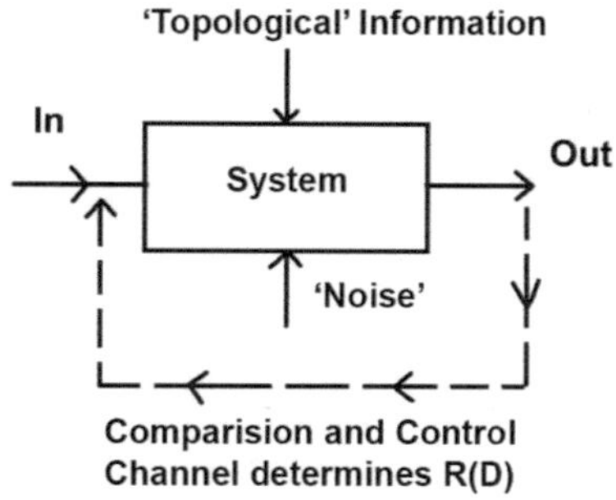

Fig. 1.1 Rate Distortion Control Theory. An inherently unstable system's topological information, as determined by the Data Rate Theorem (Nair et al. 2007), is seen as added to the effects of 'noise' to determine the minimum channel capacity $R(D)$ needed to assure transmission with average distortion D. The punctuation implied by the DRT emerges from the RDT if there is a critical maximum average distortion beyond which control fails

1.4 Toward a Model

We present a basic abstraction, abducting the standard formalism of statistical mechanics (Landau and Lifshitz 2007)—given a proper definition of 'temperature'—for a cognitive system.

Consider the full ensemble of high probability developmental trajectories available to the system, written as Y_j, $j = 1, 2, \ldots$. Each trajectory is associated with a Rate Distortion Function-defined minimum necessary channel capacity R_j for a particular maximum average scalar distortion D_j. Then, assuming some basic underlying probability model having distribution $\rho_R(C_R, x)$, where C_R is a parameter set, we can define a pseudoprobability for a meaningful 'message' Y_j sent into the system of Fig. 1.1 as

$$dP = \frac{\rho_R(C_R, R/g(Z))dR}{\int_0^\infty \rho_R(C_R, R/g(Z))dR} = \frac{\rho_R(C_R, R_j/g(Z))dR}{g(Z)} \tag{1.2}$$

since, by definition, $\rho_R(C_R, x)$ is a probability density function that integrates to one on the interval $x = 0 \to \infty$. Here, and in the following chapters, for simplicity, we only treat the continuous case. For discrete or finite-state systems, the integral is a sum and the partition function denominator becomes a more general expression $h(g(Z))$.

Again, Y_j is a particular trajectory, while the (possibly generalized) integral is over all possible 'high probability' paths available to the system.

We are implicitly imposing the Shannon-McMillan Source Coding Theorem (Cover and Thomas 2006) so that the overall set of possible system trajectories can be divided in two. These are, first, a very large set of measure zero—vanishingly low probability—that is not consistent with the underlying grammar and syntax of some basic information source, and a much smaller consistent set (Khinchin 1957).

Again, R_j is the RDT channel capacity keeping average distortion less than a given limit D_j for message Y_j. $g(Z)$ is the temperature analog from above, depending on the scalar resource rate Z. $\rho(C, x)$, for physical systems, is usually taken as the Boltzmann distribution: $\rho(x) = \exp[-x]$. We suggest that, for cognitive phenomena—from the living and institutional to the machine and composite—it is necessary to move beyond analogs with physical systems. That is, it becomes necessary explore the influence of a variety of different probability distributions, including those with 'fat tails' (Derman et al. 2016; Taleb 2020), on the dynamics of cognition/regulation stability.

Again, standard methodology from statistical physics (Landau and Lifshitz 2007) identifies the denominator of Eq. (1.2) as a partition function, giving an iterated free energy-analog F as

$$\begin{aligned} &\rho_F(C_F, F/g(Z)) \\ &\quad \equiv \int_0^\infty \rho_R(C_R, R/g(Z))\mathrm{d}R = g(Z) \\ &g(Z) = \mathrm{RootOf}\left(X - \rho_F(C_F, F/X)\right) \end{aligned} \tag{1.3}$$

where 'RootOf' means setting the expression to zero and solving for X and ρ_F is an appropriate 'free energy' distribution that must be determined case-by-case for cognitive, as opposed to simple physical, phenomena. For physical phenomena, ρ_F can often be taken as the Boltzmann distribution. Again, C_X is simply an associated parameter set.

From chemical physics (Laidler 1987), we can define a cognition rate for the system as the reaction rate analog

$$\begin{aligned} L &\equiv \frac{\int_{R_0}^\infty \rho_R(C_R, R/g(Z))\mathrm{d}R}{\int_0^\infty \rho_R(C_R, R/g(Z))\mathrm{d}R} \\ &= \frac{1}{g(Z)} \int_{R_0}^\infty \rho_R(C_R, R/g(Z))\mathrm{d}R \end{aligned} \tag{1.4}$$

R_0 is the minimum channel capacity needed to keep the average distortion below a critical value D_0 for the full system of Fig. 1.1.

The basic underlying probability model of Eq. (1.2)—via the partition function argument of Eq. (1.3)—determines system dynamics, but not system structure, and cannot be associated with a particular underlying network conformation, although these are related (Watts and Strogatz 1998; Barabasi and Albert 1999). That is, a set of markedly different networks may all be mapped on to any single given dynamic behavior pattern and, indeed, vice-versa, The same static network may display a spectrum of behaviors (Harush and Barzel 2017). We restrict focus to dynamics rather than network topology, a distinct shift from much current work.

Feynman (2000), following Bennett (1982), identifies 'information' as a form of free energy, using Bennett's clever ideal machine that turns a message directly into work. By contrast, here we are concerned here with an iterated, rather than a direct, construction.

F, as a free energy, is subject to symmetry-breaking transitions as $g(Z)$ varies (Pettini 2007). These symmetry changes, however, are not as associated with physical phase transitions as represented by standard group algebras. Such symmetry changes represent transitions from playing one 'game' to playing another. For example, a cognitive system may engage in foraging behaviors that trigger a predatory attack by another system. Then the game changes from 'foraging' to 'escape'.

Thus 'cognitive phase changes' involve shifts between equivalence classes of high probability developmental/behavioral pathways that are represented as *groupoids*, a generalization of the group concept such that a product is not necessarily defined for every possible element pair, although multiple products with multiple identity elements are defined (Brown 1987; Weinstein 1996).

Dynamics emerge via a first-order Onsager approximation akin to that of nonequilibrium thermodynamics (de Groot and Mazur 1984) in the gradient of an entropy measure constructed from the 'iterated free energy' F of Eq. (1.3). Recall from the Onsager approximation that $\partial Z/\partial t \approx \partial S/\partial Z$. The full development is

$$\begin{aligned}
S(Z) &\equiv -F(Z) + Z\mathrm{d}F(Z)/\mathrm{d}Z \\
\partial Z/\partial t &\approx \mathrm{d}S/\mathrm{d}Z = f(Z) \\
f(Z) &= Z\mathrm{d}^2F/\mathrm{d}Z^2 \\
F(Z) &= \int\int \frac{f(Z)}{Z}\mathrm{d}Z\mathrm{d}Z + C_1 Z + C_2 \\
g(Z) &= \mathrm{RootOf}\,(X - \rho_F(C_F,\ F(Z)/X))
\end{aligned} \tag{1.5}$$

An expression for $g(Z)$ follows from the second expression of Eq. (1.3). Note that the 'diffusion coefficient' has been set equal to 1 in the second expression in Eq. (1.5).

There are several important points:

- Since 'RootOf' may have complex number solutions, the temperature analog $g(Z)$ enters the realm of the 'Fisher Zeros' characterizing phase transition in physical systems (e.g., Dolan et al. 2001; Fisher 1965; Ruelle 1964, Sect. 5). Such phase transitions lie at the base of the punctuated accession to consciousness in the Baars, and related, Global Workspace Models. Indeed, the 'RootOf' construction in Eq. (1.3) actually generalizes the Lambert W-function (e.g., Yi et al. 2010; Mezo and Keady 2015).
- Information sources are not microreversible, that is, palindromes are highly improbable, e.g., 'eht' has far lower probability than 'the' in English, so that there are no 'Onsager Reciprocal Relations' in higher dimensional systems. The necessity of groupoid symmetries appears to be driven by this directed homotopy.

- Further, there will always be a delay in the rate of provision of Z, so that, in Eq. (1.5), for example, $f(Z) = \beta - \alpha Z(t)$—an exponential model having $Z(t) = (\beta/\alpha)(1 - \exp[-\alpha t])$—where $Z \rightarrow \beta/\alpha$ at a rate determined by α. Other dynamics are possible, such as the 'Arrhenius', $Z(t) = \beta \exp[-\alpha/t]$, with $f(Z) = (Z/\alpha)(\log(Z/\beta))^2$. However Z is constructed from the components $\mathscr{C}$, $\mathscr{H}$ and $\mathscr{M}$, so here, it is the scalar resource rate Z itself that counts.

1.5 Hazard Rate Tuning and Selection

For a simple physical system $\rho_F(C, x)$ is often the Boltzmann form of the exponential distribution

$$\rho_F(m, x) = m \exp[-mx]$$

with $m = 1$.

Cognitive entities are likely to be far more subtle. We suppose it possible—via selection pressure or a deliberate 'tuning' fiat—to require a particular 'hazard rate' of an embodied and regulated cognitive entity. That is, we impose a function $Q(x)$ so that

$$\frac{\rho_F(x)}{1 - \int_0^x \rho_F(u)du} = Q(x) \tag{1.6}$$

solving the integral equation for the underlying probability distribution $\rho_F(x)$.

For the exponential distribution $m \exp[-mx]$ it is easy to show that $Q(x) \equiv m$. That is, the exponential distribution has a uniform 'hazard rate' for signal detection/occurrence irrespective of x. That is, $Q(x)$ is the rate at which a new incoming signal will be recognized at load demand x. We study two cases in detail, $Q(x) = 1$, the Boltzmann distribution, and $Q(x) = 1/(1 + x)$, a detection rate that declines monotonically with increasing load.

Solving the integral relations of Eq. (1.6) for these cases gives the corresponding probability distributions as

$$\begin{aligned} \rho_F(x) &= \exp[-x] \\ \rho_F(x) &= 1/(1 + x)^2 \end{aligned} \tag{1.7}$$

The second expression is classically 'fat-tailed' (Derman et al. 2016) and does not have finite mean or variance. See Fig. 1.2, a logarithmic plot of $\rho_F(x)$ versus x.

We will be led to suggest that tuning of the hazard rate—and hence the form of the basic underlying probability distribution—may become a standard response to environmental signals or other selection pressures.

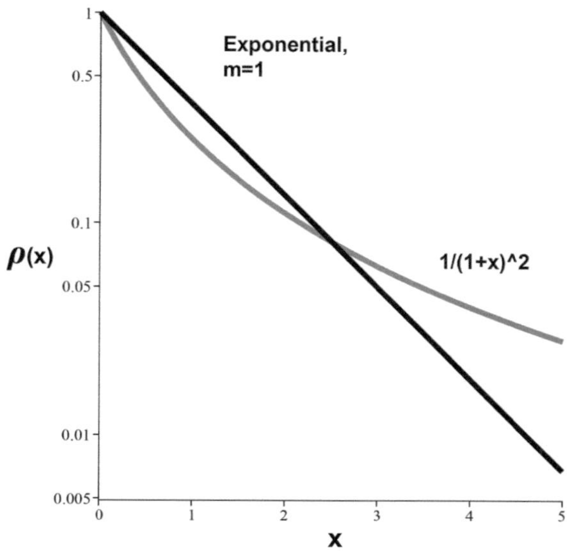

Fig. 1.2 Boltzmann and $1/(1+x)^2$ distributions on a logarithmic scale. The latter has a 'fat tail' that precludes finite mean or variance

1.6 Two Examples

We next calculate cognition rate versus 'arousal' for the two specified distribution models, the Boltzmann and $1/(1+x)^2$, first establishing that

$$g = \frac{-F}{W(n, -F)}$$
$$g_{\pm} = -F + \frac{1}{2} \pm \frac{\sqrt{1-4F}}{2} \tag{1.8}$$

where $W(n, x)$ is the Lambert W-function of order n, solving the relation

$$W(n, x)\exp[W(n, x)] = x.$$

It is real-valued only for $n = 0, -1$ and $x > -\exp[-1]$.

Notice in particular the 'Fisher Zero' phase transition conditions for the two distribution models:

$$F = \exp[-1]$$
$$F = 1/4 \tag{1.9}$$

Taking $R_0 = 1$ for the two cognition rate relations, calculation gives

$$L = \exp[-1/g]$$

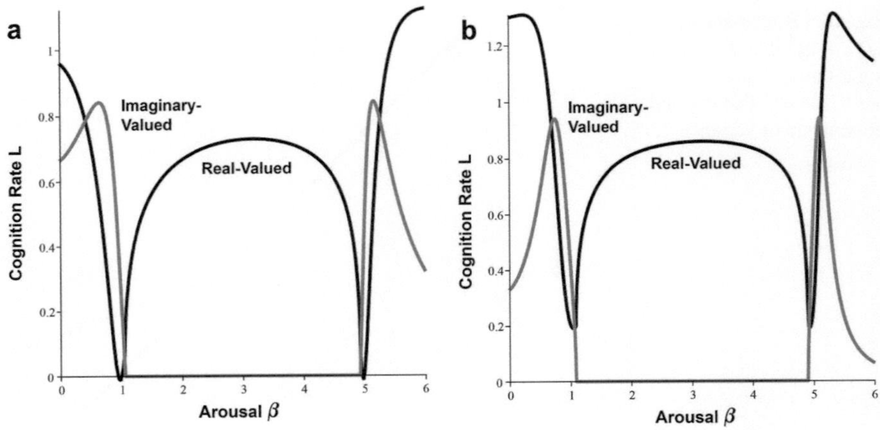

Fig. 1.3 Cognition rates versus 'arousal'. In the definition of F, $\alpha = 1$, $C_1 = -1$, $C_2 = 3$. β varies as the 'arousal' measure. **a** Exponential distribution **b** $1/(1+x)^2$ distribution. Both models are hemmed at low and high arousal by realms of 'hallucination' and 'panic' characterized by nonzero imaginary modes. The middle sections show classic inverted-U signal transduction of the Yerkes-Dodson Effect. Fisher Zero transitions punctuate the phase changes

$$L_+ = \frac{g_+}{1 + g_+} \tag{1.10}$$

where we—somewhat arbitrarily—set $\rho_R \equiv \rho_F$. Other assignments are, of course, possible and are left as exercises.

We next take an 'exponential' model for $dZ/dt = f(Z) = \beta - \alpha Z$ in Eq. (1.5), so that

$$F = \beta\,(Z\ln(Z) - Z) - \frac{\alpha^2}{Z}2 + C_1 Z + C_1 \tag{1.11}$$

Fixing $\alpha = 1,\ C_1 = -1,\ C_2 = 3$, and taking $Z \equiv \beta/\alpha$, β then varies as the index of 'arousal' in a classic Yerkes-Dodson function-versus-stress analysis (Diamond et al. 2007; Wallace and Fricchione 2024). Real-valued and imaginary components are as indicated in Fig. 1.3.

The existence of nonzero imaginary components indicates significant instability. Those to the left, at low 'arousal' β should be interpreted as 'hallucination' while those to the right, at high arousal, indicate onset of panic or thrashing dysfunction. The intermediate zones show the classic inverted-U of Yerkes-Dodson signal transduction. Fisher Zero phase transitions indicate the changes between states.

1.7 Terra Periculosa

Matters can become, of course, ever more subtle.

For the Cauchy distribution with zero offset, also 'fat-tailed' without mean or variance,

$$\rho_F(x) = \frac{2}{\pi m\left(1 + \frac{x^2}{m^2}\right)}$$

$$Q(x) = \frac{2m}{(m^2 + x^2)\left(\pi - 2\arctan\left(\frac{x}{m}\right)\right)} \tag{1.12}$$

Figure 1.4a compares the '$1/(1+x)^2$' and Cauchy distributions for $m = 1$ on a logarithmic scale, and Fig. 1.4b shows the corresponding Cauchy distribution 'hazard rate' $Q(x)$ for $m = 1$. Q first rises to a peak with increasing load, and then declines precipitously. Recall that, in our context, 'hazard rate' is actually the rate at which incoming signals can be recognized and acted upon.

The cognition 'temperature' and rate are, respectively,

$$g_+ = \frac{1 + \sqrt{1 - \pi^2 F^2}}{\pi m}$$

$$L_+ = 1 - \frac{2\arctan\left(\frac{R_0}{m g_+}\right)}{\pi} \tag{1.13}$$

again setting $\rho_R \equiv \rho_F$.

The critical values of F, analogous to Eq. (1.9), are $F = \pm 1/\pi$.

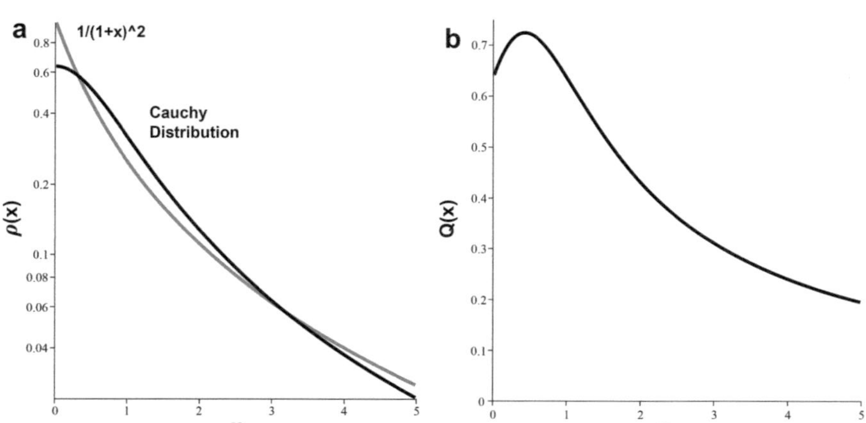

Fig. 1.4 **a** '$1/(1+x)^2$' and Cauchy distributions for $m = 1$. **b** Cauchy Distribution 'hazard rate' versus x. Unlike the previous examples, $Q(x)$ first rises to a peak, and then declines

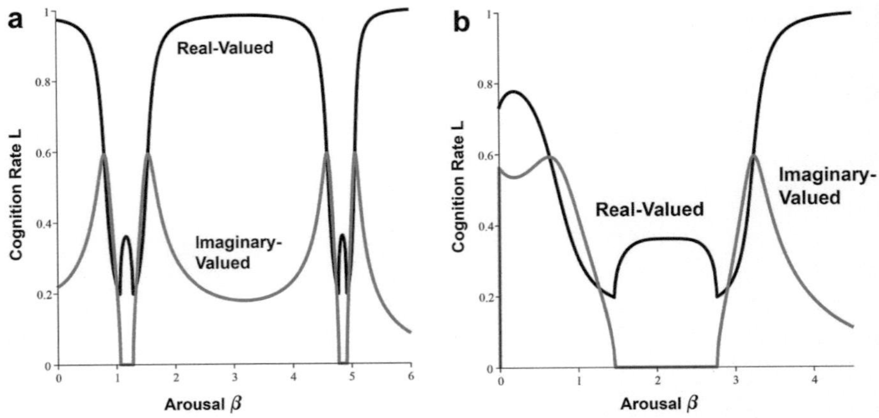

Fig. 1.5 While the Cauchy Distribution does not seem much different from the '$1/(1+x)^2$' distribution, in **a**, with boundary conditions for F being $C_1 = -1,\ C_2 = 3$, the same as in Fig. 1.3, stable cognition is limited to two very narrow 'inverted-U' spikes at low and high arousal, suggesting that inducing relatively subtle changes in a cognitive system's 'underlying probability distribution' can have profound effects on response dynamics and stability under stress. For **b**, setting $C_1 = 1,\ C_2 = 5/4$ narrowly recovers an inverted-U, essentially by merging the two purely real-valued peaks from **a**

Taking $m = R_0 = 1$, and F as in Eq. (1.11) and Fig. 1.3, the real- and imaginary-valued components of the cognition rate are shown in Figs. 1.5a and b for different boundary conditions on F. These are, respectively, for Fig. 1.5a $C_1 = -1,\ C_2 = 3$ as in Fig. 1.3, and for Fig. 1.5b, $C_1 = 1,\ C_2 = 5/4$.

Although the Cauchy Distribution does not seem, at first glance, much different from the '$1/(1+x)^2$' distribution, matters are not as they were in Fig. 1.3. Stable cognition for the conditions of Fig. 1.3, i.e., $C_1 = -1,\ C_2 = 3$ in the expression for F, is limited to two very narrow 'inverted-U' spikes at low and high arousal β. All else is lost to Dreams of Reason. The 'standard' inverted-U of the Y-D law is limited to very particular boundary conditions for F. In essence, 'tuning' the boundry conditions merges the two real-valued peaks of Fig. 1.5a.

This is not a reassuring result, suggesting that inducing relatively subtle changes in a cognitive system's 'underlying probability distribution' can have profound effects on response dynamics and stability under stress.

Again, one is struck by the possibility that a cognitive system under selection pressures of threat or affordance may resort to hazard rate tuning as a routine response.

1.8 Time Constraint and 'Hallucination'

Intelligence-as-we-know-it, in and of the real world, is constrained by resource availability, information flows, system physical structure, environmental challenge, and so on. Cognition itself requires high rates of 'free energy' of one form or another

(Wallace 2022). But a most essential resource is time-to-act under adversarial or environmental selection pressures. The 100 ms time constant required by evolutionary selection of higher animal consciousness imposes a relatively simple single tunable global workspace architecture on brain dynamics, compared with the multiple GW architectures of slower-acting immune cognition and gene expression (Wallace 2022). Time is a critical resource for the systems we study here, often addressed in terms of 'anytime algorithms' that produce a 'good enough' solution under time constraint (e.g., Zilberstein 1996). We will be particularly interested in characterizing the left-most Fisher Zero phase transitions in Fig. 1.3 as a function of time constraints.

We consider a multicomponent system acting under constraints of both the scalar Z from Sect. 1.3, and time-to-act, writing $Z = \sum_i Z_i$ and $T = \sum_i T_i$.

Reconsidering Eqs. (1.4) and (1.10), it is clear that cognition rate will be a monotonic increasing function of the 'temperature' variate g. Here, however, g is not a function of Z alone, but of a composite variable

$$X_i \equiv \frac{Z_i T_i}{\tau_i} \tag{1.14}$$

where we must introduce a characteristic time constant τ_i for dimensional considerations. We can then write a Lagrangian for an optimization argument as

$$\mathscr{L} = \sum_i g_i(X_i) + \lambda \left(Z - \sum_i Z_i \right) + \mu \left(T - \sum_i T_i \right) \tag{1.15}$$

where we consider the 'undetermined multipliers' λ, μ to be shadow price variates imposed by environmental constraints, in the sense of standard economic theory (e.g., Jin et al. 2008; Robinson 1993).

As usual, $\partial g_i/\partial Z_i = \lambda$, $\partial g_i/\partial T_i = \mu$ and $T_i = (\lambda/\mu) Z_i \equiv \kappa Z_i$, so that

$$X_i = \frac{\kappa Z_i^2}{\tau_i} \tag{1.16}$$

Equation (1.5) then becomes

$$dX/dt = X d^2 F/dX^2 = f(X) \tag{1.17}$$

for some appropriate $f(X)$

Imposing the relation $dZ_i/dt = \beta_i - \alpha_i Z_i$, and remembering that $T_i = \kappa Z_i$ so that $X_i = (\kappa Z_i^2)/\tau$, some algebra gives

$$X d^2 F/dX^2 = \frac{2\kappa}{\tau} \sqrt{\frac{\tau}{\kappa} X} \left(\beta - \alpha \sqrt{\frac{\tau}{\kappa} X} \right)$$
$$F(X) = -\alpha X^2 + \frac{8 \beta \kappa^2 \left(\frac{\tau X}{\kappa} \right)^{\frac{3}{2}}}{3\tau^2} + C_1 X + C_2 \tag{1.18}$$

where $C_1,\ C_2$ are important boundary conditions and we have dropped the index i.

We impose the model of Fig. 1.3a—the first expression of Eq. (1.8), so that the stability limit—determined by the properties of the Lambert W-functions of orders 0 and -1—becomes $F = \exp[-1]$, substituting the result of Eq. (1.16), and setting $Z_i = \beta_i/\alpha_i$.

Then the 'hallucination equation' becomes

$$\frac{\kappa^2\beta^4}{\alpha^3\tau^2} - \frac{8\beta\,\kappa^2\left(\frac{\beta^2}{\alpha^2}\right)^{\frac{3}{2}}}{3\tau^2} - \frac{C_1\kappa\,\beta^2}{\alpha^2\tau} - C_2 = -\mathrm{e}^{-1} \tag{1.19}$$

This can be directly solved for two forms of τ, giving a complicated square root result that is too long to be written here. By inspection, however, since the dominant term is $\propto \beta^4/\tau^2$—fixing α and the C_i—the β-value of the Fisher Zero hallucination point increases as $\sqrt{\tau}$.

More explicitly, in Fig. 1.6, using Eq. (1.19), we set $\alpha = 1,\ C_1 = 1,\ C_2 = -1$ and determine the hallucination value of β as a function of τ and κ for the two solution sets, as indicated. The splitting criterion between the solution forms is the sign of the shadow price ratio κ.

As the 'characteristic response time' τ increases, the hallucination point in Fig. 1.3a moves to the right $\propto \sqrt{\tau}$.

A similar calculation can be carried out for the system of Fig. 1.3b.

Higher animal consciousness, under hunting/hunted selection pressure time constraints, cannot afford large τ. By comparison, 'unbodied' Large Language Models trained on large, static data sets, have effectively infinite τ.

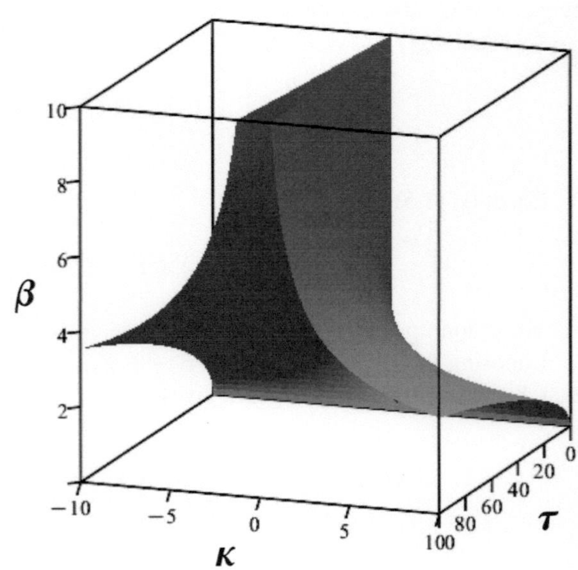

Fig. 1.6 As the 'characteristic response time' τ increases, the hallucination point in Fig. 1.3a moves to the right, $\beta \propto \sqrt{\tau}$. The sign of the shadow price ratio κ determines the solution set splitting

Dreyfus (1972, 1986), as Fjelland (2020) asserts, was right: unbodied and ‘uncultured’ machine cognition can never be ‘intelligent’.

1.9 The Dynamics of Delay

We can, of course, make matters even more complicated. Suppose, instead of the relation $dZ_i/dt = \beta_i - \alpha_i Z_i$ including only a ‘friction’ term in α_i, we add a finite fixed response delay δ in a standard manner:

$$
\begin{aligned}
dZ/dt &= \beta - \alpha Z(t - \delta) \\
Z(t) &\equiv \frac{\beta}{\alpha}(1 - \exp[St]) \\
S &= \frac{W(n,\ -\alpha\delta)}{\delta} \\
dZ/dt &= S\left(Z - \frac{\beta}{\alpha}\right)
\end{aligned}
\tag{1.20}
$$

where, again dropping the index i, the expression in $W(n, x)$—the Lambert W-function of order n—imposes complex-valued instability for $\delta\alpha > \exp[-1]$.

Then Eq. (1.18) becomes

$$
F(X) = S X^2 - \frac{8 S\beta \kappa^2 \left(\frac{\tau X}{\kappa}\right)^{\frac{3}{2}}}{3\alpha \tau^2} + C_1 X + C_2 \tag{1.21}
$$

and Eq. (1.19) transforms to

$$
\frac{W(n, -\delta\alpha)\,\kappa^2 \beta^4}{\delta\,\alpha^4 \tau^2} - \frac{8\,W(n, -\delta\alpha)\,\beta\,\kappa^2 \left(\frac{\beta^2}{\alpha^2}\right)^{\frac{3}{2}}}{3\delta\alpha\,\tau^2} + \frac{C_1 \kappa\,\beta^2}{\alpha^2 \tau} + C_2 = e^{-1} \tag{1.22}
$$

Again, there is a dominant β^2/τ term, but dynamic stability is further compromised by the real-value requirement on S, that $\alpha\delta < \exp[-1]$.

Introduction of a distributed delay, e.g.,

$$
dZ/dt = \beta - \alpha \int_0^t Z(t - \delta)\rho(\delta)d\delta \tag{1.23}
$$

where ρ is some appropriate probability distribution, leads to far more complicated relations. For example, taking $\rho(\delta) = m\exp[-m\delta]$—a simple exponential distribution—using the Laplace transform method, gives

$$Z(t) = \beta\left(\frac{\sinh\left(\frac{t\sqrt{(-4\alpha+m)m}}{2}\right)\mathrm{e}^{-\frac{mt}{2}}(-m+2\alpha)}{\sqrt{(-4\alpha+m)\,m}\,\alpha} + \frac{-\cosh\left(\frac{t\sqrt{(-4\alpha+m)m}}{2}\right)\mathrm{e}^{-\frac{mt}{2}}+1}{\alpha}\right). \tag{1.24}$$

so that $Z(t) \to \beta/\alpha$. For small m, however, the system initially displays destabilizingly high amplitude oscillations.

It is possible to generalize Eq. (1.20) to multiple delays:

$$\begin{aligned}
\mathrm{d}Z/\mathrm{d}t &= \beta - \sum_j \alpha_j Z(t-\delta_j)\\
Z(t) &= \frac{\beta}{\sum_j \alpha_j}(1-\exp[St]) \equiv \frac{\beta}{A}(1-\exp[St])\\
S &= -\sum_j \frac{\alpha_j}{\exp[S\delta_j]}
\end{aligned} \tag{1.25}$$

If $\delta_j \approx \delta$, then

$$\begin{aligned}
S &\approx \frac{-\sum_j \alpha_j}{\exp[S\delta]}\\
S &\approx \frac{W\left(n,\ -\delta\sum_j a_j\right)}{\delta}
\end{aligned} \tag{1.26}$$

where, again, $W(n,\ x)$ is the Lambert W-function of order n. This is the death of a thousand cuts.

Making the more general approximation

$$S = -\sum_j \frac{\alpha_j}{\exp[S\delta_j]} \to -\int_0^{\Delta} \alpha(\delta)\exp[-S\delta]\mathrm{d}\delta \tag{1.27}$$

and then setting $\alpha(\delta) = m\exp[-m\delta]$—the Exponential Distribution—leads to

$$\begin{aligned}
S &= \frac{m\left(\mathrm{e}^{-\Delta(S+m)}-1\right)}{S+m}\\
\Delta &= -\frac{1}{S+m}\ln\left(\frac{S^2}{m}+S+1\right)
\end{aligned} \tag{1.28}$$

The graph of the second relation, when properly rotated, directly displays $S(\Delta,\ m)$ as in Fig. 1.7a, reducing the need for possibly unstable numerical approximations. S is defined only for limited ranges of Δ and m.

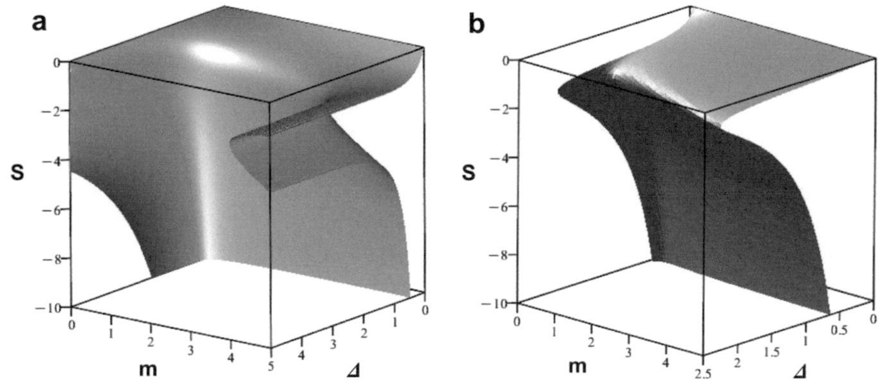

Fig. 1.7 **a** Rotated solution set for $S(\Delta, m)$ in the second expression of Eq. (1.28). S is real-valued only over a very limited ranges of m and the delay index Δ. **b** Numerical solution for the distribution $m/(m+x)^2$. Each distribution, of course, has it's own details, but the general forms are similar

Other possible distributions for $\alpha(\delta)$, like the Gamma, the extension of the second form of Eq. (1.7) as $m/(m+x)^2$, shown in Fig. 1.7b, and the Cauchy—even being fat-tailed—provide recognizable analogs to Fig. 1.7a, albeit requiring numerical methods. In Fig. 1.7b we use the *implicitplot3d* function of the computer algebra program Maple 2024.

A stochastic extension of Eq. (1.20) is of some interest, setting

$$dZ_t = S\left(Z_t - \frac{\beta}{\alpha}\right)dt + \sigma Z_t dB_t \tag{1.29}$$

where dB_t is taken as ordinary Brownian white noise.

Using the Ito Chain Rule on Z^2 (Protter 2005), the variance is expressed as

$$\mathrm{Var}(Z) = \left(\frac{S\frac{\beta}{\alpha}}{\frac{\sigma^2}{2} + S}\right)^2 - \left(\frac{\beta}{\alpha}\right)^2 \tag{1.30}$$

Again, the system becomes grossly unstable if $\delta\alpha > \exp[-1]$ since S is then complex-valued.

Note, however, that real-valued S is always negative so that Eq. (1.30) explodes as $\sigma^2/2 \to |S|$.

The result of Eq. (1.30) applies to the multiple-delays situation of Eqs. (1.25)–(1.27) by replacing α with A, as properly defined for a distributed system.

1.10 Discussion

Successful high-level, real-world, real-time cognition is always paired with effective regulation and, as we argue in Sects. 1.5–1.7, possibly with selective hazard rate tuning as well. The immune system, while tasked with routine cellular maintenance, malignancy suppression, and pathogen containment, is not allowed unrestricted self-attack. In vivo protein folding is closely monitored and 'chaperoned'. During the gestation and growth of higher animals, gene expression is meticulously regulated. The human stream of consciousness is confined by social and cultural riverbanks to realms useful to the social group as well as to the self. The instruments of governance apply law enforcement to constrain both individual behavior and the practices of component social institutions. Simple biology. Simple sociology. Simple political science.

Individualistic, atomistic, disconnected Tech Bro 'AGI' will be no more useful than current driverless cars are in Midtown traffic at rush hour. Or near a primary school at the end of the day.

Much of such regulation is necessarily enfolded by, and held within, general ideas of 'embodiment' as we have described it, albeit with somewhat simplistic models. More comprehensive approaches, similarly based on the asymptotic limit theorems of information and control theories, will likely lead to the development of powerful statistical tools useful in the design, construction, and ongoing monitoring of real-world, real-time cognitive systems, up to and including useful, if limited, machine entities. Such entities will not, however, be what is currently envisioned as 'AGI' (Fjelland 2020).

Western Tech Bro atomistic cognitive machines will not function well confronting the 'Clausewitz landscapes' of fog, friction, and adversarial intent that characterize much of real-world experience. As history has shown, Darwinian selection challenges the culture-bound ideologies of even the most powerful empires.

References

Barabasi, A., R. Albert, 1999, Emergence of scaling in random networks, Science 286:509–512.

Bennett, C.H., 1982, The thermodynamics of computation, International Journal of Theoretical Physics 21:905–940.

Brown, R., 1987, From groups to groupoids: a brief survey, Bulletin of the London Mathematical Society, 19:113–134.

Cover, T., J. Thomas, 2006, Elements of Information Theory, Second Edition, Wiley, New York.

de Groot, S., P. Mazur, 1984, Non-Equilibrium Thermodynamics, Dover, New York.

Derman, E., M. Miller, D. Park, 2016, The Volatility Smile, Wiley, New York.

Diamond, D., A. Campbell, C. Park, J. Halonen, P. Zoladz, 2007, The Temporal Dynamics Model of Emotional Memory Processing Neural Plasticity, https://doi.org/10.1155/2007/60803.

Dolan, B., W. Janke, D. Johnston, M. Stathakopoulos, 2001, Thin Fisher zeros, Journal of Physics A 34:6211–6223.

Dreyfus, H., 1972, What Computers Can't Do, Harper and Row, New York, NY.

Dreyfus, H., S. Dreyfus, 1986, Mind Over Machine, Blackwell, Basil.

Feynman, R., 2000, Lectures on Computation, Westview Press, New York.
Fisher, M., 1965 Lectures in Theoretical Physics vol. 7, University of Colorado Press, Boulder.
Fjelland, R., 2020, Why general artificial intelligence will not be realized, Humanities and Social Sciences Communications, 7:10, https://doi.org/10.1057/s41599-020-0494-4.
Harush, U., B. Barzel, 2017, Dynamic patterns of information flow in complex networks, Nature Communications 8:2181
Heine, S., 2001, Self as cultural product: an examination of East Asian and North American selves, Journal of Personality 69:881–906.
Jin, H., Z. Hu, X. Zhou, 2008, A convex stochastic optimization problem arising from portfolio selection, Mathematical Finance 18:171–183.
Khinchin, A., 1957, The Mathematical Foundations of Information Theory, Dover, New York.
Laidler, K., 1987, Chemical Kinetics, Third Edition, Harper and Row, New York.
Landau, L., E. Lifshitz, 2007, Statistical Physics, 3rd Ed. Part 1, Elsevier, New York.
Markus, H., S. Kitayama, 1991, Culture and the self-implications for cognition, emotion, and motivation, Psychological Review 98:224–253.
Masuda, T., R. Nisbett, 2006, Culture and change blindness, Cognitive Science 30:381–399.
Mezo I., G. Keady, 2015, Some physical applications of generalized Lambert functions, arXiv:1505.01555v2 [math.CA] 22 Jun 2015.
Nair, G., F. Fagnani, S. Zampieri, R. Evans, 2007, Feedback control under data rate constraints: an overview, Proceedings of the IEEE 95, 108138.
Nisbett, R., K. Peng, C. Incheol, A. Norenzayan, 2001, Culture and systems of thought: holistic vs. analytic cognition, Psychological Review 108:291–310.
Nisbett, R., Y. Miyamoto, 2005, The influence of culture: holistic versus analytic perception, Trends in Cognitive Science 9:467–473.
Pettini, M., 2007, Geometry and Topology in Hamiltonian Dynamics and Statistical Mechanics, Springer, New York.
Protter, P., 2005, Stochastic Integration and Dierential Equations,Second Edition, Springer, New York.
Robinson, S., 1993, Shadow prices for measures of e ectiveness II: Generalmodel, Operations Research 41:536–548.
Roy, N., et al., 2021, From machine learning to robotics: Challenges and opportunities for embodied intelligence, arXiv:2110.15245v1 [cs.RO] 28 Oct 2021.
Ruelle, D., 1964, Cluster property of the correlation functions of classical gases, Reviews of Modern Physics, April, 580–584.
Simons, D., C. Chabris, 1999, Gorillas in our midst: sustained inattentional blindness for dynamic events, Perception 28:1059–1074.
Stratton, P., 2024, Convolutionary, evolutionary, and revolutionary: What's next for brains, bodies, and AI? Cognitive Computation 16:2420–2427.
Taleb, N. N. (ed.), 2020, Statistical Consequences of Fat Tails: Real world preasymptotics, epistemology, and applications, STEM Academic Press.
Wallace, R., 2007, Culture and inattentional blindness. Journal of Theoretical Biology 245:378–390.
Wallace, R., 2018, Canonical Instabilities of Autonomous Vehicle Systems: The unsettling reality behind the dreams of greed, Springer, New York.
Wallace, R., 2021. How AI founders on adversarial landscapes of fog and friction, Journal of Defense Modeling and Simulation https://doi.org/10.1177/1548512920962227.
Wallace, R., 2022, Consciousness, Cognition, and Crosstalk: The evolutionary exaptation of nonergodic groupoid symmetry-breaking, Springer, New York.
Wallace, R., D. Wallace, 2016, Gene Expression and its Discontents: The social production of chronic disease, Second Edition, Springer, New York.
Wallace, R., G. Fricchione, 2024, Stress-induced failure of embodied cognition: a general model, BioSystems, https://www.sciencedirect.com/science/article/abs/pii/S0303264724000789.
Watts, D., S. Strogatz, 1998, Collective dynamics of small world networks, Nature 393:440–442.

Weinstein, A., 1996, Groupoids: unifying internal and external symmetry, Notices of the American Mathematical Association, 43:744752.
Yi, S., P.W. Nelson and A.G. Ulsoy, 2010, Time-Delay Systems: Analysis and Control Using the Lambert W Function, World Scientific, New Jersey.
Zilberstein, S., 1996, Using anytime algorithms in intelligent systems, AI Magazine, 17:73–83.

Chapter 2
Dreams of Reason: 'AGI', 'Machine Consciousness' and Foundational Embodiment

Abstract A surprisingly direct model based on Rate Distortion Control Theory finds that 'intelligent systems' having large characteristic embodiment-and-response times relative to those of embedding selection pressures cannot constrain distortion between what is wanted and what is done, or even ensure basic stability. Systems that are not adequately embodied—or, indeed, not embodied at all and hence under permanent sensory deprivation—will have very limited realms of utility. Current venture-capital research streams related to Artificial General Intelligence and Machine Consciousness do not properly focus on the foundational necessity of hyperrapid embodiment under selection dynamics. Indeed, via their static training sets, contemporary large language models are effectively without any form of embodiment. The resulting Dreams of Reason cannot be sustained or endured.

> ***Look, we don't have popular crackpot theories for nuclear fusion, but we do for consciousness. I'm betting fusion power is up-and-running a lot sooner than AGI is useful.***
>
> —Anonymous (2024).

> ***The whole endeavour of the consciousness studies community is absurd – they are in pursuit of a chimera. They misunderstand the nature of consciousness. The conception of consciousness which they have is incoherent. The questions they are asking don't make sense. They have to go back to the drawing board and start all over again. It's literally a total waste of time.***
>
> —Hacker (2010).

> ...***Heidegger argued that we can only have a concept of a hammer or a chair because we belong to a culture where we grow up and are able to handle these objects. Dreyfus... therefore thought that computers, who have no body, no childhood and no cultural practice, could not acquire intelligence at all***....
>
> —Fjelland (2020).

R. Wallace, *Hallucination and Panic in Autonomous Systems*, Studies in Computational Intelligence 1217, https://doi.org/10.1007/978-3-031-95207-4_2

2.1 Introduction

The current venture-capital-driven surge in Artificial Intelligence research and production carries the burden of a grossly simplistic ‘Tech Bro’ conception of cognition and consciousness that is debrided of sophisticated evolutionary—and even simple biological—perspective. In sum, you can’t make, remake, replicate, or ‘improve’ what you don’t really understand.

Evolutionary biology, in large measure, is destiny, and rapid and responsive embodiment arising from elaborate and path-dependent evolutionary process is the central point of intelligence as-we-know-it in higher animals, including humans. No hyperrapid embodiment, involving close feedback coupling with a rich embedding world, no real ‘intelligence’ (Dreyfus 1972; Dreyfus and Dreyfus 1986). Simple, direct, compelling, and diametrically opposite to present bizarrely-narrow Tech Bro venture capital thinking.

There are no benefits to be accrued from an intellectual gilded age: being very rich does not make one automatically correct in complex matters of science and technology, naive social construction, echo-chamber self-indulgence, and fanboy media coverage notwithstanding.

And so, on with it: how to actually make an ‘intelligent machine’ for real-world enterprise.

Although it may seem odd to say, we already know how to do this. Humans have been building large, effective intelligent machines throughout their evolutionary history. These machines are our elaborate social structures, economic systems, and formal institutions. Contending with each other and with their environments, they benefit—or suffer—from evolutionary selection (Wallace 2020, 2023). Here, too, rapid feedback from, and response to, the embedding ‘signal space’—a special kind of embodiment—is central.

It is, surprisingly, not difficult to model essential features of such embodiments using Rate Distortion Control Theory.

We begin with a characteristic simplification.

2.2 Scalarizing Resource Rates

We envision a multicomponent system embedded in, and interacting with, a larger world, but operating under constraints of both time and resource rates.

These essential resource rates are taken to be:

- The rate at which internal signals can be sent between individual system subcomponents, C.
- The rate at which ‘sensory/intelligence’ information about the outside world is available, H.
- The rate at which metabolic free energy and/or other material or ‘materiel’ resources are made available, M.

These rates can often be compounded into a scalar $Z(C, H, M)$ that accounts for a large fraction of the system variance, analogous to principal component analysis. Most simply, this can be done as $Z = C \times H \times M$. Typically, however, one must contend with an (at least) 3 by 3 matrix $\mathbf{Z}$ that includes cross-influence terms, something like the familiar correlation matrix.

An n-dimensional square matrix $\mathbf{X}$ typically has n scalar invariants r_j, $j = 1 \ldots n$ under appropriate transformations that can be found using the standard decomposition

$$p(\gamma) = \det(\mathbf{X} - \gamma \mathbf{I}) = (-1)^n \gamma^n + (-1)^{n-1} r_1 \gamma^{n-1} + \cdots - r_{n-1}\gamma + r_n \quad (2.1)$$

$\mathbf{I}$ is the n-dimensional identity matrix, det is the determinant, and γ is a real-valued parameter. Under many conditions, it is often possible to project the preponderance of system variability down on to a single scalar function of these invariants, which we write as $Z = Z(r_1, \ldots, r_n)$. The method is analogous to a principal component analysis of a correlation matrix. If more than one parameter is required, as Wallace (2021) indicates, matters become rapidly more mathematically complicated.

2.3 Rate Distortion Control Theory

Sims (2016) describes the central matter as follows:

> Perception is the act of extracting meaning from noisy and uncertain sensory signals, and in the process choosing what information to transmit and what to discard. Once perceived, perceptual memory is the act of sending a message to your future self. The fundamentally communicative nature of perception and memory suggests the relevance of information theory to the study of perceptual processing. However, for biological information processing systems, it is not enough to merely transmit information. Rather, the goal of perceptual processing must be to help the organism achieve goals. This suggests a utilitarian perspective on human perception. Rate-distortion theory... represents the mathematical framework combining these two disciplines: information theory and decision theory.

The Rate Distortion Theorem (RDT) characterizes the channel capacity needed for message transmission under conditions of noise. There will be, for that particular channel, assuming some scalar measure of average distortion, D, between what is sent and what is received, a minimum necessary channel capacity $R(D)$ (CoverThomas 2006). In a sense, the theorem asks what is the 'best' channel for transmission of a message with the least possible average distortion. $R(D)$ can be defined for nonergodic information sources via a limit argument based on the ergodic decomposition of a nonergodic source into a 'sum' of ergodic sources.

For the systems we study, however, R will be determined as some monotonic increasing function of the composite measure Z defined above, *in the context of the time and resource constraints.*

The RDT can be reconfigured in a control theory context by viewing systematic and highly-structured 'topological information' imposed by an embedding 'roadway'

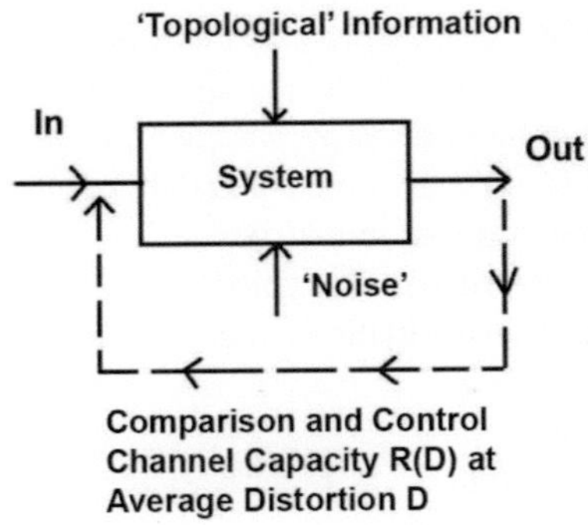

Fig. 2.1 Envisioning system control through the lens of the Rate Distortion Theorem. The system-of-interest sends a 'control signal' into an environment, and then measures the distortion between what has been 'sent' and how the environment responds. An environment's systematic 'topological information' is seen as added to the effects of noise to determine the minimum channel capacity $R(D)$ needed to assure transmission with average distortion D. Punctuation can emerge if there is a critical maximum average distortion for the system. Systems may degrade more gracefully or can have even more complex dynamic patterns

environment as another form of noise, adding to the average distortion D (e.g., Futrell (2023) and references therein). See Nair et al. (2007) for a more complete discussion of 'topological information' in the sense of the Data Rate Theorem. That is, the system-of-interest sends a 'message' trying to control the embedding environment, and estimates the distortion between what has been demanded and what has been observed. See Fig. 2.1. 'Simple' phase transition-like punctuation can occur if there is a critical maximum allowed average distortion. Other systems may, however, degrade gracefully, or, as we will show below, have even more complicated patterns of stability and instability.

2.4 A 'Simple' Model

Our 'intelligent' system is made up of n linked subcomponents, each competing for both resources and time under constraints of basic availability and under the external effects of 'environmental shadow prices' in the sense of standard economic theory (e.g., Jin et al. 2008; Robinson 1993).

We impose a 'worst case' Gaussian channel (CoverThomas 2006) and incorporate both 'topological information' and 'noise' under the Shannon-Hartley (SH) description of channel capacity R_i *for each subcomponent of the full system.* We further impose resource and time constraints as $Z = \sum_i T_i$ and $T = \sum_i T_i$. Then

$$R_i = B_i \ln\left(1 + \frac{Z_i T_i/\tau_i}{\sigma_i^2}\right) \tag{2.2}$$

B_i is the subsystem 'bandwidth'. The term $Z_i T_i / \tau_i$ is the magnitude of the 'signal' under the SH model as parameterized by a characteristic time constant of the subcomponent studied, τ_i. This is important, since, while Z_i is a rate, T_i is a time duration that must be normalized by τ_i for dimensional consistency. σ^2 is seen as incorporating both 'noise' and the oppositional 'topological information'—in the sense of the Data Rate Theorem—being imposed by the embedding environmental 'roadway'.

For an optimization calculation, a standard Lagrangian can be constructed as

$$L = \sum_i R_i + \lambda \left(Z - \sum_i Z_i \right) + \mu \left(T - \sum_i T_i \right) \tag{2.3}$$

Recalling the optimization conditions $dL/dZ_i = 0$, $dL/dT_i = 0$, solving for Z_i and T_i, plugging these values into Eq. (2.2)—dropping the index i—there are two solutions,

$$R_{\pm} = B \ln \left(\frac{B \left(B \pm \sqrt{B^2 - 4\lambda\mu\,\sigma^2\tau} \right)}{2\lambda\mu\,\sigma^2\tau} \right) \tag{2.4}$$

There is an evident stability condition in the bandwidth B for each subcomponent,

$$B^2 - 4\lambda\mu\sigma^2\tau > 0 \tag{2.5}$$

so that the minimum bandwidth required for stability increases as $\sqrt{\tau}$.

Indeed, 'effective time constants' $\mathcal{T}_i$—and minimum necessary bandwidths—can be defined in terms of Eq. (2.5) as

$$\begin{aligned} \mathcal{T}_i &\equiv 4\lambda\mu\sigma_i^2\tau_i \\ B_{min} &= \sqrt{\mathcal{T}} \end{aligned} \tag{2.6}$$

Figure 2.2 sets $B = \lambda = \mu = \sigma = 1$ and explores the effect of variation in the characteristic time constant τ for the physically-realizable R_+ solution(s).

The real-valued component of R_+, given a zero-valued imaginary component, declines rapidly and monotonically with increasing τ until the stability criteria of Eqs. (2.5) and (2.6) are violated.

Note that, by virtue of the log-term in Eq. (2.4), a negative shadow price product, i.e., $\lambda\mu < 0$, will always generate a nonzero imaginary component, imposing system instability for all values of τ and σ.

To reiterate, even fixing τ at a low value, a sufficient positive term $\lambda\mu\sigma^2$—imposing a synergism of environmental shadow price constraint and 'noise'—serves to raise the 'effective response time' to destabilizing levels, in consonance with the military strategist John Boyd's famous 'OODA Loop' ideology, i.e. 'get inside your opponent's decision loop' (Osinga 2007; Wallace 2020, 2021b). This is illustrated in

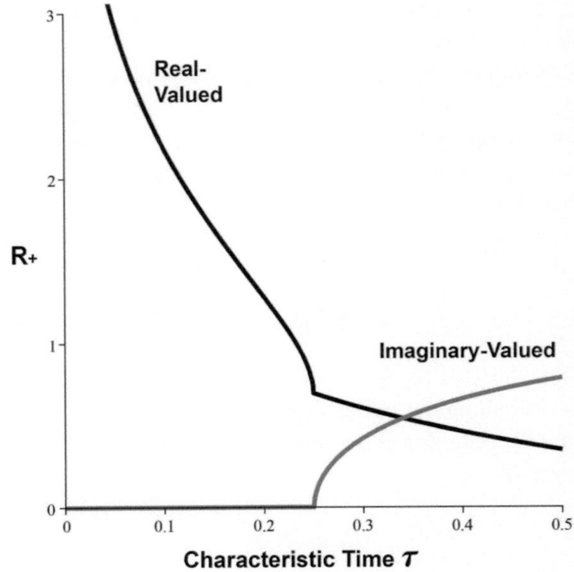

Fig. 2.2 The physically-realizable R_+ solution set from Eq. (2.4). Here, $B = \lambda = \mu = \sigma = 1$, and the RDT channel capacity R_+ is determined as a function of τ. The real-valued component of R_+, given a zero imaginary-valued component, declines rapidly and monotonically with increasing τ until the stability criterion is violated

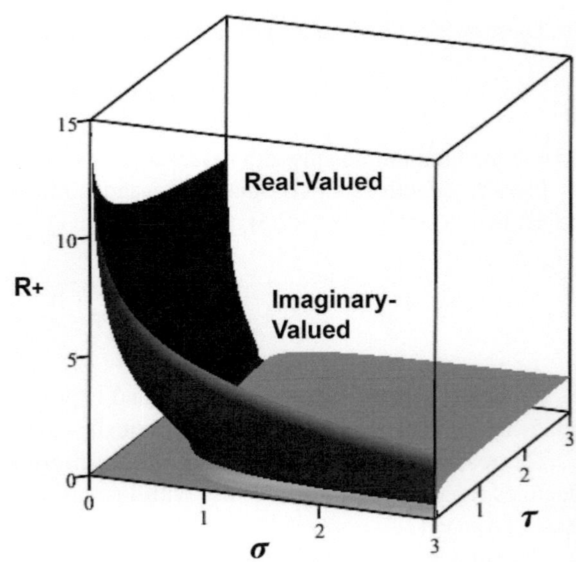

Fig. 2.3 An 'OODA loop' version of Fig. 2.2, allowing the 'noise' σ to vary along with τ. Other parameters are as in Fig. 2.2. Rising 'noise'—the fog-and-friction of organized conflict—demands ever-shorter characteristic system times to avoid onset of thrashing instabilities

Fig. 2.3, again taking B, λ and μ as in Fig. 2.2, allowing the 'noise' σ to co-vary with τ. Rising 'noise'—the fog-and-friction of organized conflict—demands ever-shorter characteristic system times to avoid onset of thrashing instabilities.

2.5 Discussion

Slow systems, having large 'effective response times' $\mathscr{T} = 4\lambda\mu\sigma^2\tau$ in comparison with the short characteristic times of embedding selection pressures, cannot limit distortion between what is wanted or needed and what is done, or even ensure basic dynamic stability via an argument that parallels the Data Rate Theorem (Nair et al. 2007). Systems that are not adequately embodied to address the rapidly changing selection pressures of the 'roadway' on which they operate—or, indeed, are not embodied at all and hence under permanent and inherent sensory deprivation—will have very limited realms of actual utility.

Current venture-capital-driven research funding streams related to AI, 'AGI' and 'machine consciousness,' do not sufficiently focus on the necessity of hyperrapid embodiment under the stunning variety of rapidly-changing real-world selection pressures. Indeed, via their static training sets, current large language models are effectively without embodiment.

Translation of these considerations into realms of institutional cognition and dynamics under selection is straightforward. One thinks of the Wehrmacht on the Eastern Front, and of the US in Korea after Inchon, in Vietnam, the occupation of Iraq, and, of course, the UK, the USSR, and the USA in Afghanistan, the graveyard of empires.

As has been said, The Real World is not a Game of Go (Wallace 2018).

All in all, as Fjelland (2020) argues, Dreyfus (1972) and Dreyfus and Dreyfus (1986) was right.

There is, of course, some considerable understanding of these matters among reality-oriented technologists. Etwaru (2024), for example, writes

> One of the key arguments against the feasibility of AGI is the importance of embodiment in the development of intelligence. Human intelligence is deeply rooted in our physical experiences and interactions with the world. This concept, known as embodied cognition, posits that our cognitive processes are shaped by our physical bodies and the environment in which we operate.
>
> Artificial systems, lacking a physical body and the rich sensory experiences that come with it, face significant challenges in developing the kind of understanding humans possess. Without the ability to physically interact with the world, an AGI would be deprived of the context and grounding essential for true comprehension and learning....

This being said, sufficient 'embodiment' might well be provided AI entities acting within such severely limited real-world sandboxes as air-to-air combat, permitting a degree of stable utility. After all, (some) armies do sufficiently well on 'Clausewitz landscapes' of fog, friction, and deadly adversarial intent.

But even such machines, of limited embodiment and hence of limited capability, will be nothing like the masterful AGI of mythical SkyNet, or even like a tiger hunting in the night.

References

Cover, T., J. Thomas, 2006, Elements of Information Theory, Second Edition, Wiley, New York.

Dreyfus, H., 1972, What computers can't do. Harper and Row, New York.

Dreyfus, H., S. Dreyfus, 1986, Mind Over Machine, Blackwell, Basil.

Etwaru, R., 2024, There could never be an artificial general intelligence, Forbes, July 1, https://www.forbes.com/councils/forbestechcouncil/2024/07/01/there-could-never-be-an-artificial-general-intelligence/.

Fjelland, R., 2020, Why general artificial intelligence will not be realized, Humanities and Social Sciences Communications, https://doi.org/10.1057/s41599-020-0494-4.

Futrell, R., 2023, Information-theoretic principles in incremental language production, PNAS 120:e2220593120, https://doi.org/10.1073/pnas.2220593120.

Hacker, P., 2010, How not to Argue, The Philosophers' Magazine 51(4) 23–32 https://doi.org/10.5840/tpm2010517.

Jin, H., Z. Hu, X. Zhou, 2008, A convex stochastic optimization problem arising from portfolio selection, Mathematical Finance 18:171–183.

Nair, G., F. Fagnani, S. Zampieri, R. Evans, 2007, Feedback control under data rate constraints: an overview, Proceedings of the IEEE 95:108–138.

Osinga, F., 2007, Science, strategy and war: the strategic theory of John Boyd, Routledge, London.

Robinson, S., 1993, Shadow prices for measures of effectiveness II: General model, Operations Research 41:536–548.

Sims, C., 2016, Rate-distortion theory and human perception, Cognition 152:181–198.

Wallace, R., 2018, Carl von Clausewitz, the Fog-of-War, and the AI Revolution: The real world is not a game of Go, Springer, New York.

Wallace, R., 2020, Cognitive Dynamics on Clausewitz Landscapes: The control and directed evolution of organized conflict, Springer, New York.

Wallace, R., 2021a, How AI founders on adversarial landscapes of fog and friction, Journal of Defense Modeling and Simulation, https://doi.org/10.1177/1548512920962227.

Wallace, R., 2021b, On Maskirovka: the dynamics of delay in threat recognition, Journal of Defense Modeling and Simulation, https://doi.org/10.1177/15485129211061688.

Wallace, R., 2023, Essays on the Extended Evolutionary Synthesis: Formalizations and extensions, Springer, New York.

Chapter 3
The Regulation and Stabilization of Embodied Cognition

Abstract The draconian regulation required by the inherent instability of cognitive phenomena—gene expression, immune function, cancer suppression, wound healing, animal consciousness, machine intelligence, network stabilization, institutional cognition, and their many and varied composites—can be viewed through the lens of the asymptotic limit theorems of both information and control theories. Here, we explore the dynamics and sometimes highly punctuated failures of the regulation of cognition under increasing 'noise'. The approach parallels, and indeed generalizes, the Data Rate Theorem of control theory, extending the theorem's requirement of a minimum channel capacity necessary for stabilization of an inherently unstable system. Various models are explored across different basic underlying probability distributions characteristic of the system under study, and across different hierarchical scales, finding the addition of adaptive—learned—regulation greatly extends the reach of innate—e.g., AI-driven or otherwise pre-programmed—regulation. This work points toward the construction of new statistical tools for the analysis of observational and empirical data across a wide spectrum of inherent pathologies and adversarial challenges afflicting a wide variety of cognitive processes.

3.1 Introduction

As a consequence of the inherent unruliness of biological, institutional, and machine cognition—and their many composites—under real-world stress, there are surprisingly few foundational 'principles' available to the intrepid explorer of such landscapes.

That most often recognized is, of course, some form of Darwinian 'Reproduction with Variation and Selection' underlying all path-dependent evolutionary phenomena (Gould 2002).

A second is the assertion by Maturana and Varela (1980) that the living state in particular is cognitive at every scale and level of organization. Machine intelligence and institutional function follow increasingly closely.

A further principle was articulated by Dretske (1994) and Atlan and Cohen (1998), who recognized that all cognitive process—most particularly, for Dretske, those

R. Wallace, *Hallucination and Panic in Autonomous Systems*, Studies in Computational Intelligence 1217, https://doi.org/10.1007/978-3-031-95207-4_3

that bear particular meaning and are semantic—must be constrained by necessary conditions imposed by the asymptotic limit theorems of information theory. The basic statement that emerges from the work of Atlan and Cohen is that cognition requires choice and choice inherently reduces uncertainty. The reduction of uncertainty implies the existence of an information source 'dual' to the cognitive process studied. The inference is direct, intuitive, unambiguous, and finesses controversies regarding 'representation'.

Wallace (2023a, Chap. 4) expands the Dretske/Atlan-Cohen insights by noting that, for the living state in particular, cognition must be paired with regulation, since embodied cognition is inherently unstable as it must act in a real world that imposes it's own 'topological information' on most processes of cognition and action. Again, real-world cognition demands stabilization by imposition of control information at a rate greater than the cognitive system is burdened by extrinsic topological information. The resulting dynamics can be viewed from perspectives of both the Data Rate Theorem (DRT) of control theory (Nair et al. 2007), and the Rate Distortion Theorem (RDT) of information theory (Cover and Thomas 2006).

Thus gene expression must be highly regulated during early development for successful growth, and failure of regulation is a burden of aging. The immune system must be constrained from self-attack, while directed to engage in vigorous cancer identification and suppression, pathogen resistance, and routine tissue maintenance. For higher animals, the stream of consciousness must be constrained by cultural and social riverbanks to realms useful for the individual animal, the local social grouping, and the full species. Coordinated social behaviors are constrained by cultural norms. Institutions must respond to extrinsic topological information and noise from the perspective of internal 'doctrine' in a large sense. Machine cognition in its various forms that is faced with twisting roadway dynamics falls under much the same rubric.

We attempt a formal model, focusing on the added affordance provided by an 'adaptive', in addition to an 'innate', regulatory capacity.

First, we review some methodological boilerplate.

3.2 The Rate Distortion Theorem

French et al. (1999) explored the dynamics of a neural network model of the famous Yerkes-Dodson mechanism (Wallace and Fricchione 2024), interpreting that mechanism in terms of a necessary real-time data compression that eventually fails as task complexity increases. Data compression inherently involves matters central to information theory, in addition to those of control theory. And there is a formal context: the Rate Distortion Theorem (Cover and Thomas 2006).

Recall that information theory has three classic asymptotic limit theorems (Cover and Thomas 2006; Khinchin 1957):

- The Shannon Coding Theorem. This states that, for a fixed—i.e., stationary—transmission channel, a message recoded so as to be 'typical' with respect to the

probabilities of that channel, can be transmitted without error at a channel capacity rate C characteristic of the channel. One can, vice versa, argue for a tuning theorem variant of the coding theorem in which a transmitting channel is tuned to be made typical with respect to the message being sent (Wallace et al. (2022a), p. 83), so that, formally, the channel is, in a sense, 'transmitted by the message' at a dual channel capacity C^*.
- The Shannon-McMillan or Source Coding Theorem. This states—for our purposes—that messages transmitted by an information source along a stationary channel can be divided into sets, a very small one congruent with a characteristic grammar and syntax, and a much larger one of vanishingly small probability that is not congruent. For stationary (in time), ergodic sources, where long-time averages are cross-sectional averages, the splitting criterion dividing the two sets is given by the classic Shannon uncertainty. For nonergodic sources, which are likely to predominate in biological and ecological circumstances, matters are more complex, requiring a 'splitting criterion' that must be associated with each individual high-probability message (Khinchin 1957).
- The Rate Distortion Theorem (RDT). This addresses the question of message transmission under conditions of noise for a particular information channel. There will be, for that channel, given a particular scalar measure of average distortion, D, between what is sent and what is received, a minimum necessary channel capacity $R(D)$. In a sense, the theorem asks what is the 'best' channel for transmission of a message with the least possible average distortion. $R(D)$ can be defined for nonergodic information sources via a limit argument based on the ergodic decomposition of a nonergodic source into a 'sum' of ergodic sources.

The RDT can be reconfigured in a control theory context, if we envision a system's topological information from the DRT as 'simply' another form of noise, adding to the average distortion D. See Fig. 3.1. The punctuation implied by the DRT (Nair et al. 2007) emerges from this model if there is a critical maximum average distortion that characterizes the system. Other systems may degrade more gracefully, or, as we will show below, have even more complicated patterns of punctuation that, in effect, generalize the Data Rate Theorem.

Here, we expand perspectives on the dynamics of biocognition/bioregulation dyads across rates of arousal, across the possibly manifold set of basic underlying probability distributions that characterize such dyads at different scales, levels of organization, and indeed, across various intrinsic patterns of arousal.

3.3 Scalarizing Essential Resource Rates

The RDT holds that, under a given scalar measure of average distortion D between a sequence of signals that has been sent and the sequence that has actually been received in message transmission—measuring the difference between what was ordered and what was observed—there is a minimum necessary channel capacity $R(D)$ in the

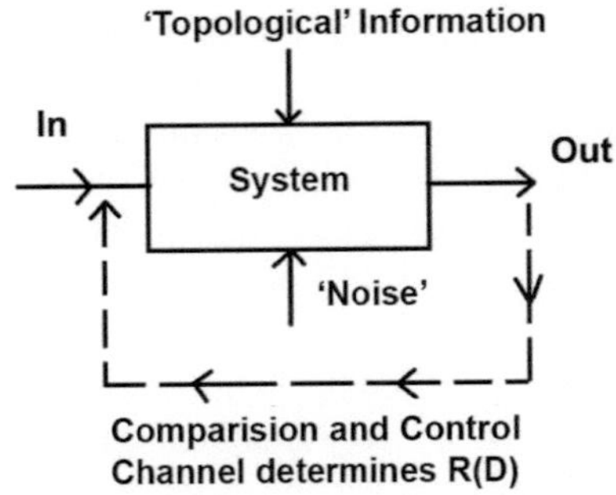

Fig. 3.1 Reinterpretation of a control system in terms of the Rate Distortion Theorem. An unstable system's 'topological information' (Nair et al. 2007) is seen as added to the effects of noise to determine the minimum channel capacity $R(D)$ needed to assure transmission with average distortion D. The punctuation implied by the DRT emerges from the RDT if there is a critical maximum average distortion that characterizes the system

classic information theory sense that is, in turn, determined by the rate at which a set of essential resources, indexed by some particular scalar measure Z, can be provided to the system 'sending the message' in the presence of noise and an opposing rate of topological information characteristic of the inherent instability of the control system under study.

The Rate Distortion Function (RDF) $R(D)$ is necessarily convex in D, so that $d^2R/dD^2 \geq 0$ (Cover and Thomas 2006; Effros et al. 1994; Shields et al. 1978). For a nonergodic process in which the cross-sectional mean is not the same as the time-series mean, the RDF can still be defined as the average across the RDF's of the ergodic components of that process, and is thus convex in D.

The relations between the minimum necessary channel capacity $R(D)$ and the rates of any essential resources are, however, likely to be subtle.

A 'simple' scalar resource rate index—say Z—which we adopt as a first approximation, must be composed of a minimum of three interacting components:

- The rate at which subcomponents of the system of interest can communicate with each other, defined by a channel capacity $\mathcal{C}$.
- The rate at which 'sensory' information is available from the embedding environment, associated with a channel capacity $\mathcal{H}$.
- The rate $\mathcal{M}$ at which 'metabolic' or other free energy 'materiel' resources can be provided to a subsystem of a full entity, organism, institution, machine, and so on.

These rates must be compounded into an appropriate scalar measure $Z(\mathcal{C}, \mathcal{H}, \mathcal{M})$. Most simply, this might perhaps be taken as their product, the sum of their logs, or some other such.

Following Wallace and Fricchione (2024), 'Z' will most likely be a 3 by 3 matrix, including necessary interaction crossterms. An n-dimensional matrix $\mathbf{Z}$ has n scalar invariants r_i under appropriate transformations that are determined by the characteristic equation

$$\det[\mathbf{Z} - \gamma\mathbf{I}] = \sum_{i=1}^{n} (-1)^{i-1} \gamma^{i-1} r_i \tag{3.1}$$

I is the n-dimensional identity matrix, det the determinant, and γ a real-valued parameter. r_1 is the matrix trace and r_n the matrix determinant. Given these n scalar invariants, it will often be possible to construct Z as a scalar function of them, analogous to the principal component analysis of a correlation matrix.

Wallace (2021), however, provides an example in which two such indices are minimally necessary, a matter requiring sophisticated Lie Group methods.

3.4 The Fundamental Model

Feynman (2000) and Bennett (1982) argue that information is a form of free energy, not an 'entropy', in spite of information theory's Shannon uncertainty taking the same mathematical form as entropy for a simple—indeed, simplistic—ergodic system. Feynman (2000) illustrates this equivalence using a simple ideal machine to convert information from a message into useful work.

We next apply something much like the standard formalism of statistical mechanics—given a proper definition of 'temperature'—for a cognitive, as opposed to a 'simply' physical, system. This approach leads, however, to an important diversion from the dynamics implied by simple physical theory-based arguments.

We consider the full ensemble of high probability developmental trajectories available to the system, writing these as $Y_j,\ j = 1, 2, \ldots$. Each trajectory is associated with a Rate Distortion Function-defined minimum necessary channel capacity R_j for a particular maximum average scalar distortion D_j. Then, assuming some basic underlying probability model having distribution $\rho(c, x)$, where c is a parameter set, we can define a pseudoprobability for a meaningful 'message' Y_j sent into the system of Fig. 3.1 as

$$dP = \frac{\rho(c, R/g(Z))dR}{\int_0^\infty \rho(c, R/g(Z))dR} = \frac{\rho(c, R_j/g(Z))dR}{g(Z)} \tag{3.2}$$

since ρ is a probability density that integrates to unit value.

To reiterate, Y_j is a particular trajectory, while the (possibly generalized) integral is over all possible 'high probability' paths available to the system.

We are implicitly imposing the Shannon-McMillan Source Coding Theorem so that the overall set of possible system trajectories can be divided in two. These are, first, a very large set of measure zero—vanishingly low probability—that is not consistent with the underlying grammar and syntax of some basic information source, and a much smaller consistent set (Khinchin 1957).

Again, R_j is the RDT channel capacity keeping average distortion less than a given limit D_j for message Y_j. $g(Z)$ is a yet to be determined temperature analog depending on the scalar resource rate Z. $\rho(c, x)$, for physical systems, is usually taken as the Boltzmann distribution: $\rho(x) = \exp[-x]$. We suggest that, for cognitive phenomena—from the living and institutional to the machine and composite—it is necessary to move beyond analogs with physical systems. That is, it becomes necessary to explore the influence of a variety of different probability distributions, including those with 'fat tails' (Derman et al. 2016; Taleb 2020), on the dynamics of cognition/regulation stability.

Standard methodology from statistical physics (Landau and Lifshitz 2007) identifies the denominator of Eq. (3.2) as a *partition function*. This allows definition of an *iterated* free energy-analog F as

$$\rho(c, F/g(Z)) \equiv \int_0^\infty \rho(c, R/g(Z))\mathrm{d}R = g(Z) \tag{3.3}$$

Again adapting a standard argument, now from chemical physics (Laidler 1987), we can define a cognition rate for the system as a reaction rate analog

$$\begin{aligned} L &\equiv \frac{\int_{R_0}^\infty \rho(c, R/g(Z))\mathrm{d}R}{\int_0^\infty \rho(c, R/g(Z))\mathrm{d}R} \\ &= \frac{1}{g(Z)} \int_{R_0}^\infty \rho(c, R/g(Z))\mathrm{d}R \end{aligned} \tag{3.4}$$

R_0 is the minimum channel capacity needed to keep the average distortion below a critical value D_0 for the full system of Fig. 3.1.

The basic underlying probability model of Eq. (3.2)—via Eq. (3.4)—determines system dynamics, but not system structure, and cannot be associated with a particular underlying network conformation, although these are related (Watts and Strogatz 1998; Barabasi and Albert 1999). That is, a set of markedly different networks may all be mapped on to any single given dynamic behavior pattern and, indeed, vice-versa, The same static network may display a spectrum of behaviors (Harush and Barzel 2017). We restrict focus to dynamics rather than network topology, a distinct shift from much current work.

Next, abducting a canonical first-order approximation from nonequilibrium thermodynamics (de Groot and Mazur 1984), we can now define a 'real' entropy—as opposed to a 'Shannon uncertainty' that is basically Feynman's (2000) free energy—from the *iterated* free energy F by taking the standard Legendre Transform (Landau and Lifshitz 2007), so that

$$S \equiv -F + Z\mathrm{d}F/\mathrm{d}Z \tag{3.5}$$

A—indeed, the usual—first-order Onsager approximation from nonequilibrium thermodynamics is then (de Groot and Mazur 1984)

$$\partial Z/\partial t \approx dS/dZ = ZdF^2/dZ^2 \tag{3.6}$$

The scalar diffusion coefficient has been set equal to one.

In the Mathematical Appendix we explore a 'canonical' expansion of the Legendre Transform of Eq. (3.5), analogous to building 'higher order' regression models.

For multidimensional systems there can be no 'Onsager reciprocal relations' between components of the diffusion tensor, since information sources are not microreversible. For example, in English the term 'it' does not have the same probability as 'ti'.

The next step is to impose a burden of stochastic noise on the system, done in classic manner via the stochastic differential equation (Protter 2005)

$$dZ_t = Z_t d^2F/dZ^2 dt + \sigma Z_t dB_t \tag{3.7}$$

where the last term is simple volatility in the white noise dB_t, with the scalar value σ characterizing the magnitude of the effect (Derman et al. 2016; Taleb 2020).

A basic assumption is that, to survive, the cognitive system of interest must have stability in the variance of Z, so that the nonequilibrium steady state condition $\langle dZ_t^2 \rangle = 0$ is maintained. Application of the Ito Chain Rule to Z_t^2 based on Eq. (3.7) finds (Protter 2005)

$$\begin{aligned} \langle dZ_t^2 \rangle &= 2Z^2 d^2F/dZ^2 + \sigma^2 Z^2 = 0 \\ F(Z) &= -\frac{\sigma^2 Z^2}{4} + C_1 Z + C_2 \end{aligned} \tag{3.8}$$

Determination of $F(Z)$ from Eq. (3.8), in turn, fully characterizes the temperature analog $g(Z)$ via Eq. (3.3). We can then calculate the cognition rate from Eq. (3.4).

See the chapter Mathematical Appendix for a brief summary and review of material on stochastic differential equations.

3.5 'Innate' and 'Adaptive' Regulation

The next step appears deceptively simple. We break the regulatory channel into two distinct components, having respective channel capacities R—'innate'—and aR—'adaptive'—that, in first approximation, combine into a single total $(1+a)R$. After some elementary calculus, the relations of interest become

$$\begin{aligned} \rho(c, F/g) &= \frac{g}{1+a} \\ L &= \int_{R_0(1+a)/g}^{\infty} \rho(c, u)du \end{aligned} \tag{3.9}$$

The first expression, based on defining F according to Eq. (3.8) so that the variance of Z is confined to a nonequilibrium steady state, gives $g(Z)$. The second, based on the calculation for $g(Z)$, determines the cognition rate of the combined innate/adaptive system as a function of a, σ and Z.

Examples are both surprising and contradictory.

The Boltzmann Distribution

For the Boltzmann Distribution, taking $C_1 = -1,\ C_2 = 3$ in the expression of F for Eq. (3.8),

$$
\begin{aligned}
\exp[-F/g] &= \frac{g}{1+a} \\
g &= \frac{-F}{W\left(n,\ \frac{-F}{1+a}\right)} \\
L &= \exp[-R_0(1+a)/g]
\end{aligned}
\tag{3.10}
$$

$W(n,\ x)$ is the Lambert W-function of order n that satisfies the relation

$$
W(n,\ x)\exp[W(n,\ x)] = x
$$

It is real-valued only for $n = 0,\ -1$ in the range $x > -\exp[-1]$.

Fixing $\sigma = 1$ and letting both a and Z vary gives the cognition rate pattern of Fig. 3.2. There are two essential points. The first is the decline in the hallucination realm in Z with increasing a. The second is the decline in what is effectively the minimum cognition rate necessary for second-order stability with increasing a.

The Cauchy Distribution

The 'fat-tailed' Cauchy Distribution, having undefined mean and variance, gives a different pattern. The basic relations, again taking F as above, are

$$
\begin{aligned}
\frac{2\,s}{\pi\left(\frac{F^2}{g^2} + s^2\right)} &= \frac{g}{a+1} \\
g &= \frac{1}{\pi s}\left(1 + a + \sqrt{a^2 + 2a + 1 - \pi^2 F^2}\right) \\
L &= 1 - \frac{2}{\pi}\arctan\left(\frac{R_0(a+1)}{gs}\right)
\end{aligned}
\tag{3.11}
$$

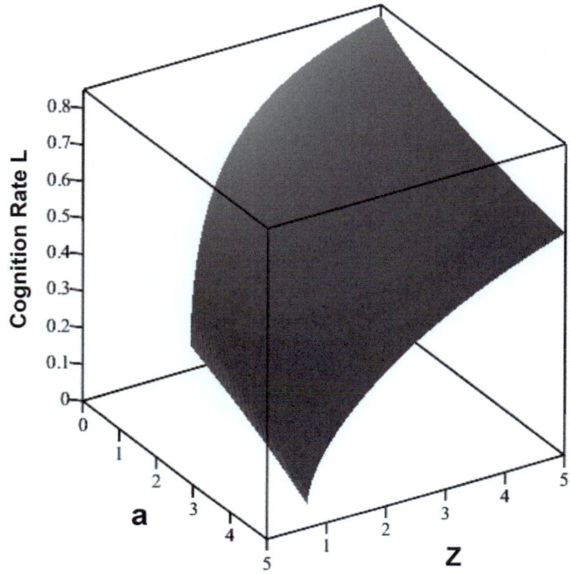

Fig. 3.2 Note the significant declines in the hallucination realm in Z and the necessary cognition rate for second-order stability with increasing a

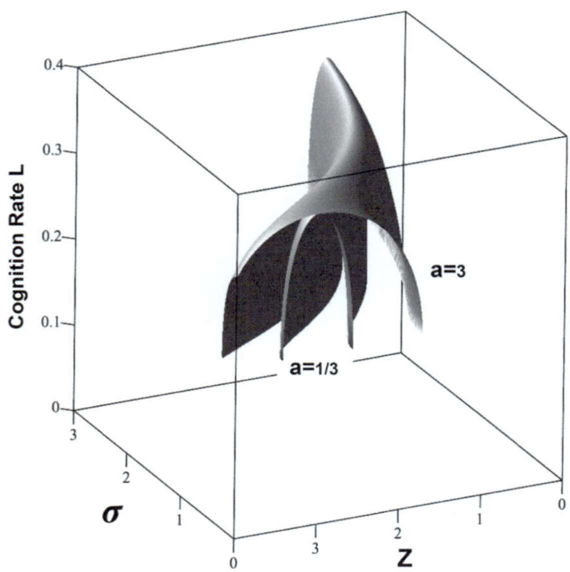

Fig. 3.3 Here, using F as above, with $C_1 = -1, C_2 = 3$ and $s = 1$, rising 'noise' σ drastically narrows an agent's ecological niche regardless of a, the relative index of adaptive regulation. It is striking, however, that for the Cauchy Distribution, a higher level of adaptive regulation greatly broadens the niche at low noise

Taking $a = 1/3,\ 3,\ s = 1$, gives Fig. 3.3, much different from the Boltzmann example in two important ways. Here, rising noise σ drastically narrows the agent's ecological niche regardless of a, the relative index of adaptive regulation. It is striking, however, that for the Cauchy Distribution, a higher level of adaptive regulation greatly broadens the niche at low noise.

The Gamma Distribution

The full Gamma Distribution model is

$$\frac{\left(\frac{F}{g}\right)^{k-1} \mathrm{e}^{-\frac{F}{g\theta}}}{\Gamma(k)\,\theta^k} = \frac{g}{1+a}$$

$$g = -\frac{F}{k\theta\,\mathrm{W}\left(-\frac{\left(\frac{F\,\theta^k\,\Gamma(k)}{1+a}\right)^{\frac{1}{k}}}{k\theta}\right)}$$

$$L = \frac{\left(2\,g^2 + (2a+2)\,g + (1+a)^2\right)\mathrm{e}^{\frac{-1-a}{g}}}{2\,g^2} \tag{3.12}$$

noting that the mean and variance are, respectively, $k\theta$ and $k\theta^2$. Again, $W(x)$ is the zero-order Lambert W-function, and the full expression for cognition rate L is graced by a Whittaker M-function and many terms. Here, for L, we have set $k = 3,\ \theta = 1$.

This particular model representation is shown in Fig. 3.4, again with $a = 1/3,\ 3$, and F as above. Although, here, the two niches are not nested, again niche width collapses with increasing σ, and the width for $a = 3$ is much greater than for $a = 1/3$.

Many other distributions, e.g., the lognormal, give results recognizably similar to Figs. 3.3 and 3.4. Here, Fig. 3.2—the exponential/Boltzmann distribution—seems a distinct outlier.

Matters rapidly become more complex under hierarchical structuring.

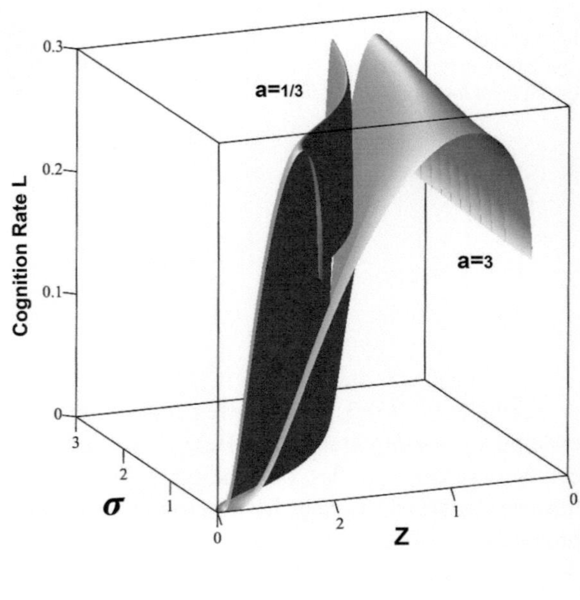

Fig. 3.4 Cognition rates for the Gamma Distribution models, setting $k = 3,\ \theta = 1$, and again taking $a = 1/3,\ 3$ and F as above. Although the two niches are not nested in this example, again niche width implodes with rising σ, and the width for $a = 3$ is much greater than for $a = 1/3$

3.6 Hierarchy Under Environmental Constraint

We next examine a networked multicomponent cognition/regulation system where subcomponents operate under the 'Onsager' relations of Eq. (3.6)

$$\partial Z_i/\partial t \approx \mathrm{d}S_i/\mathrm{d}Z_i = Z_i \mathrm{d}^2 F_i/\mathrm{d}Z_i^2 \tag{3.13}$$

The full system has $i = 1, 2, \ldots, m$ cooperating linked units that must be individually supplied with resources at rates Z_i under overall constraints of resources and time as $\sum_i Z_i = Z$, $\sum_i T_i = T$. We seek to optimize—minimize—the overall distortion under these constraints.

There are, of course, quite complicated general approaches to optimization (Nocedal and Wright 2006), but here simple Lagrangian optimization across the sum of the distortions D_i provides a sufficient formal base:

$$\mathscr{L} = \sum_i D_i + \lambda \left(Z - \sum_i Z_i \right) + \mu \left(T - \sum_i T_i \right)$$
$$\partial \mathscr{L}/\partial Z = \lambda, \quad \partial \mathscr{L}/\partial T = \mu, \quad \partial D_i/\partial Z_i = \lambda, \quad \partial D_i/\partial T_i = \mu \tag{3.14}$$

In economic theory, the undetermined multipliers λ and μ represent 'shadow prices' imposed on system dynamics by environmental constraints, in a large sense (Jin et al. 2008; Robinson 1993).

Again, simple Lagrangian optimization is enough to outline the basic mechanism.

We are now concerned with the *overall* relations $\partial\mathscr{L}/\partial Z = \lambda$ and $\partial\mathscr{L}/\partial T = \mu$. Heuristically,

$$\begin{aligned} \partial Z/\partial T &= \mu/\lambda \equiv Z d^2 \mathscr{F}/dZ^2 \\ \mathscr{F} &= \frac{\mu}{\lambda} Z\,(\ln(Z) - 1) + C_1 Z + C_2 \end{aligned} \tag{3.15}$$

where $\mathscr{F}$ is a new—global—free energy construct on which there are imposed the distribution constraints of Eq. (3.9).

The Mathematical Appendix explores a possible stochastic extension of this result.

We next use the Boltzmann Distribution relations of Eq. (3.10), but now founded on $\mathscr{F}$ of Eq. (3.15), not on Eq. (3.8). Here, $\lambda = \mu = R_0 = 1$, with $C_1 = -1$, $C_2 = 3$ in Eq. (3.15). Figure 3.5 shows the cognition rate calculations for $a = 1/3, 3$.

The Boltzmann Distribution over the full hierarchical system now manifests a more-or-less standard inverted-U signal transduction Yerkes-Dodson effect (Wallace and Fricchione 2024). Again, however, niche width increases with the degree of 'adaptive' regulatory capacity.

These patterns are profoundly affected by the shadow price ratio, as shown in Fig. 3.6, where μ/λ varies between ± 1. Again, $R_0 = 1$. Certain combinations of shadow price ratio and Z constrict niche width.

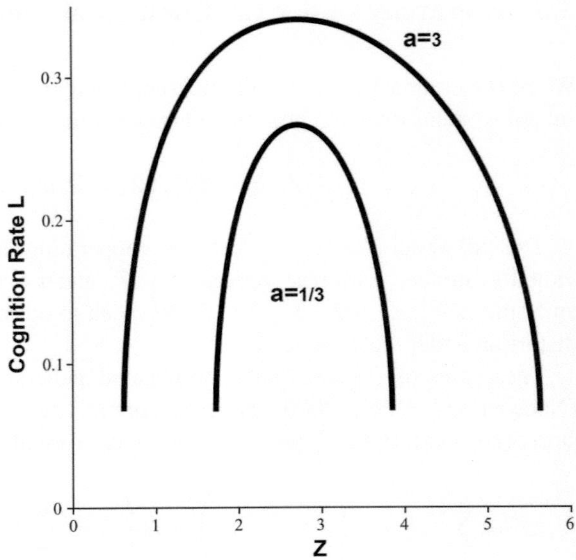

Fig. 3.5 Hierarchical cognition rate based on the Boltzmann Distribution. Here $\lambda = \mu = R_0 = 1$ for L, while $C_1 = -1$, $C_2 = 3$ in $\mathscr{F}$. The full system now manifests a standard inverted-U signal transduction

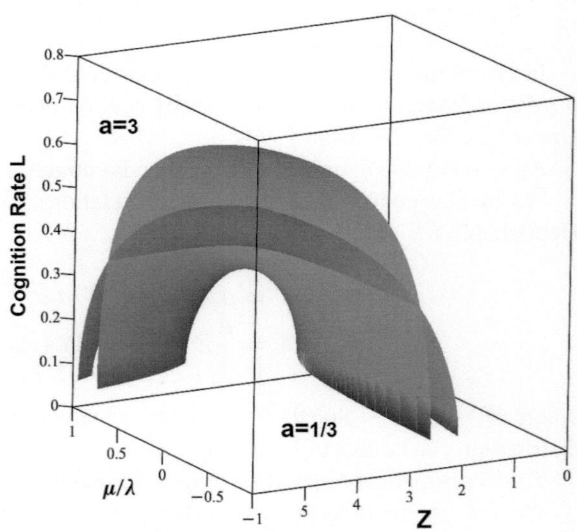

Fig. 3.6 Extension of Fig. 3.5, allowing the shadow price ratio μ/λ to vary. Again, $R_0 = 1$. Certain combinations of shadow price ratio and resource rate constrict niche width

The hierarchical cognition rate calculation can be carried out for the Cauchy Distribution example, again using Eq. (3.15). Here, $s = R_0 = 1$ and $\mathscr{F}$ is as in Fig. 3.6, again taking $a = 1/3, 3$. While the cognition rate surfaces are more complicated than for the Boltzmann example, the inverted-U reappears. Again, the niche for $a = 3$ is broader than for $a = 1/3$.

Similarly, the Gamma Distribution cognition rate example can be recalculated for the hierarchical model, again setting $k = 3$, $\theta = 1$, $R_0 = 1$, and taking $\mathscr{F}$ as in Figs. 3.5, 3.6 and 3.7. This is shown in Fig. 3.8, again with $a = 1/3, 3$ and $C_1 =$

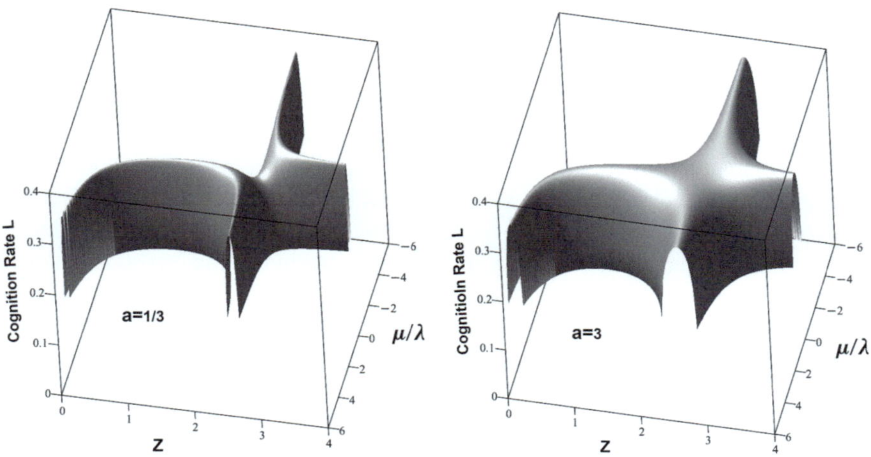

Fig. 3.7 The calculation of Fig. 3.6 can be reprized for the Cauchy Distribution, taking $s = R_0 = 1$. $\mathscr{F}$ is as in Fig. 3.6. While the cognition rate surfaces are more complex than for the Boltzmann example, inverted-U signal transduction reappears, with the niche for $a = 3$ generally broader than for $a = 1/3$

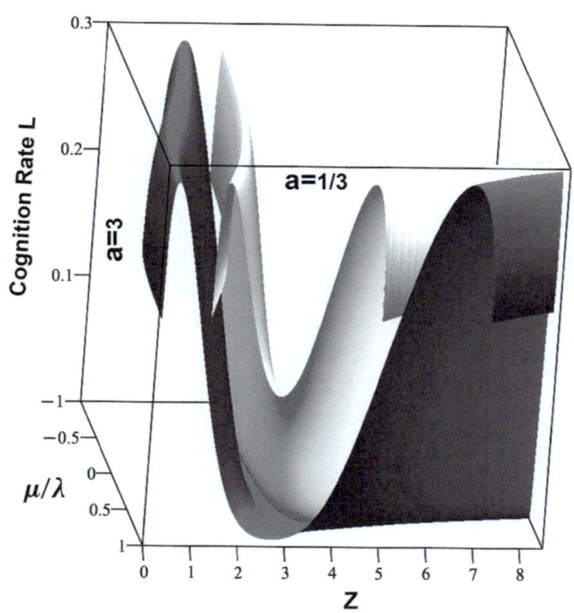

Fig. 3.8 The Gamma Distribution model in the hierarchical case again shows the inverted-U signal transduction, with the cognition rate broadened and stabilized by the larger adaptive index a

-1, $C_2 = 3$. The inverted-U signal transduction reemerges, with the higher a-value both broader and less compacted by decline in shadow price ratio.

The interested reader can explore these matters for other probability distributions. Inverted-U signal transduction—the Yerkes-Dodson effect—is ubiquitous under the hierarchical structure defined by Eq. (3.15).

3.7 Conclusions

As argued elsewhere (Wallace 2023b; Butlin et al. 2023), constructing a recognizable 'conscious machine'—of dubious stability and utility beyond attracting venture capital—seems only a matter of cleverly assembling current off-the-shelf commodity electronic components. This is, in large measure, because higher animal consciousness, to meet 100 ms time constant requirements, is a grossly simplified version of mechanisms exapted by evolutionary process from the inevitable crosstalk between cognitive subcomponents of organisms (Wallace 2012, 2022b). Wound healing, gene expression, immune maintenance-and-response, and so on, by virtue of their extended time horizons, can entertain multiple, interacting, dynamic global workspaces. A 100 ms neural system cannot, even with a metabolic temperature-equivalent an order of magnitude higher than muscle tissue. For organic systems, high speed requires simple structures under metabolic constraints: classic tradeoff.

A fast multiple workspace cognitive process of any nature—effectively, a high speed artificial immune system—must, like the organic varieties, be heavily regulated, if only to prevent autoimmune self-attack. In addition, of course, there is the problem of simultaneous, coordinated effort in multiple targeting. These matters already confront and confound military air defense systems and, indeed, 'ordinary' civilian air (and other) traffic control, albeit in an often grossly simplified environment.

We have deconstructed and parsed, in a sense, an evolutionary solution to some of these ambiguities: the innate and adaptive immunities. A high speed artificial immune system's 'innate AI' would, in this portrait, have been carefully trained on preexisting observational or experimental data. The adaptive system—via a de-facto 'embodiment'—would, in some contrast, be expected to learn on-the-job, as it were, to continually fine-tune and update response/control mechanisms under the influence of a rapid exchange of signals with the embedding environment. This is likely to be a good deal harder to engineer well than a single global-workspace conscious machine. The payoff, however, would be considerable.

In addition to such venture capital matters, we have explored the more general dynamics—and sometimes highly punctuated failures—of the regulation of cognition under increasing noise of any sort. The approach parallels, and indeed generalizes, the Data Rate Theorem of control theory, extending the theorem's requirement of a minimum channel capacity necessary for stabilization of an inherently unstable system. This work has studied various models across different basic underlying probability distributions characteristic of systems of possible interest, and across different hierarchical scales. We have explored in detail how the addition of adaptive—learned—regulation greatly extends the reach of innate—e.g., genetic, AI-driven, doctrinal, or otherwise pre-programmed—regulation.

This work, at the very least, indicates how to construct a new set of robust statistical tools for the analysis of observational and empirical data across a wide spectrum of pathologies and adversarial challenges afflicting a wide variety of important cognitive phenomena.

Mathematical Appendix

A Stochastic Version of the Hierarchical Model

Recall Eq. (3.15):

$$\begin{aligned}\partial Z/\partial T &= \mu/\lambda \equiv K = Zd^2\mathscr{F}/dZ^2 \\ \mathscr{F} &= KZ(\ln(Z) - 1) + C_1 Z + C_2\end{aligned}$$

An 'intuitively direct' stochastic extension seeks the nonequilibrium steady state (ness) solution $\langle d\mathscr{F}_t \rangle = 0$ using the base SDE

$$dZ_t = K dt + \sigma Z_t dB_t \tag{3.16}$$

where, as in Eq. (3.7), dB_t is Brownian white noise and the last term represents 'ordinary volatility'.

Application of the Ito Chain Rule finds the equation for the ness of $\mathscr{F}$ to be

$$K\,(K\,(\ln(Z) - 1) + K + C_1) + \frac{KZ\sigma^2}{2} = 0 \tag{3.17}$$

This has the nonequilibrium steady state solution

$$Z_{nss} = \frac{2\,KW\left(\frac{\sigma^2 e^{-\frac{C_1}{K}}}{2\,K}\right)}{\sigma^2} \tag{3.18}$$

where, again, $W(x)$ is the zero-order Lambert W-function, real-valued only for $x > -\exp[-1]$. The expression for Z can then be plugged in to various models to study stochastic effects under hierarchical shadow price constraints.

Different values of the boundary condition C_1 provide markedly different patterns of critical behavior across the set $\{K, \sigma\}$.

For fixed C_1 and $K > 0$, however, the rising level of 'noise' σ can have profound effect, as in Fig. 3.9. Here, $K = 1$, $C = -1$ and σ varies as indicated. Sufficient 'noise' drives Z_{nss} below critical value.

Expanding the Legendre Transform

Equations (3.5) and (3.6) are only the first of a large class of possible regression models analogs. Dimensionally-consistent extensions without cross-terms can be written as

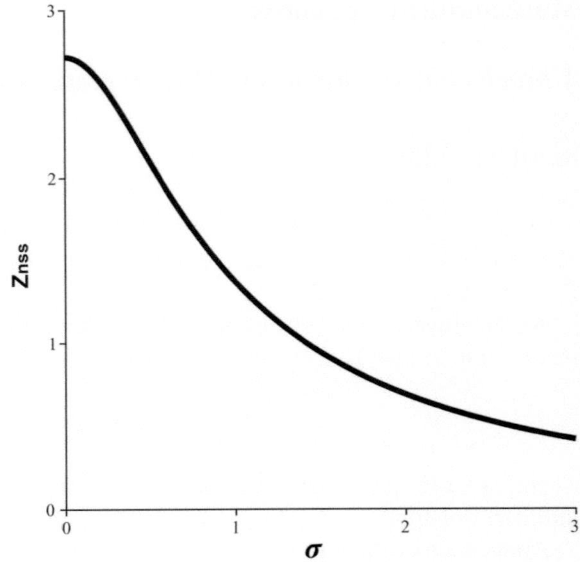

Fig. 3.9 Nonequilibrium steady state value of Z from Eq. (3.18), setting $C_1 = -1,\ K = 1$. For shadow price ratio $K > 0$, as σ increases, Z will fall below critical level

$$S \equiv -F(Z) + ZdF/dZ + \sum_{n=2}^{\infty} \epsilon_n Z^n d^n F/dZ^n$$

$$\partial Z/\partial t = \mu dS/dZ + \sum_{n=2}^{\infty} \mu_n^{n-1} d^n S/dZ^n \tag{3.19}$$

As Jackson et al. (2017) put it,

> [T]he Legendre transform can be viewed as mapping the coefficients of one formal power series into the coefficients of another formal power series. Here, the term 'formal' does not express 'mathematically nonrigorous', as it often does in the physics literature. Instead, the term 'formal power series' is here a technical mathematical term, meaning a power series in indeterminates. Formal power series are not functions. A priori, formal power series merely obey the axioms of a ring and questions of convergence do not arise.

In contrast with physical models, the ϵ and μ spectra—and possible series crossterms—raise fundamental empirical questions typical of all statistical models, analogous to fitting regression equations to observational or experimental data.

It is worth following out a simple example. Suppose we define

$$S_{\pm} \equiv -F(Z) + ZdF/dZ \pm Z^2 d^2 F/dZ^2$$
$$dZ/dt \approx dS_{\pm}/dZ$$
$$dZ_t \approx [dS_{\pm}/dZ]dt + \sigma Z_t dB_t \tag{3.20}$$

where we are expanding Eqs. (3.5)–(3.7) in terms of the first expression of Eq. (3.20), again setting the 'diffusion coefficient' equal to 1.

Imposing the second-order stability condition $\langle dZ_t^2 \rangle = 0$ allows, after a little algebra and the Ito Chain Rule, derivations of $F_\pm$ that generalize the second expression of Eq. (3.8),

$$
\begin{aligned}
F_+(Z) &= -\frac{\sigma^2 Z^2}{12} + \frac{C_1}{2Z} + C_2 Z + C_3 \\
F_-(Z) &= \frac{\sigma^2 Z^2}{4} + C_1 \left(Z \ln(Z) - Z \right) + C_2 Z + C_3
\end{aligned}
\tag{3.21}
$$

For an underlying Boltzmann Distribution model, the temperature analog is given by the relation

$$
\begin{aligned}
\exp[-F/g] &= g \\
g &= \frac{-F}{W(n, -F)}
\end{aligned}
\tag{3.22}
$$

where $W(n, x)$ is the Lambert W-function of orders $n = 0, -1$. The stability condition for real-valued g is then

$$
-F_\pm > -\exp[-1] \tag{3.23}
$$

determining the location of 'Fisher Zero' phase transitions representing onset of 'hallucination' and 'panic' modes.

The relations $F_\pm = \exp[-1]$ can be solved by inspection for σ^2 in Eq. (3.21). Figure 3.10 sets $C_1 = -1$, $C_2 = C_3 = 1$ and explores the solution sets $\{Z, \sigma\}$ for $F_\pm$. For F_-, the 'hallucination' mode kicks in at a very low value of 'noise' σ, while F_+ displays a bifurcated pattern across increasing σ.

A more complete treatment defines the entropy-analog as

$$
S \equiv -F(Z) + Z dF/dZ + \frac{Z^2}{n} d^2 F/dZ^2 \tag{3.24}
$$

so that, for subsequent algebraic simplicity, the absolute value of the second-order term declines with increasing n. This may be either positive or negative. The stability limit of Eq. (3.23) then becomes

$$
F(Z) = \frac{C_1 Z^{-n}}{(n+1)\,n} - \frac{\sigma^2 n\, Z^2}{4n+8} + C_2 Z + C_3 = e^{-1} \approx 0.3679 \tag{3.25}
$$

Solving for σ, and setting the C_i as in Fig. 3.10, we take $n = \pm 3/2, \pm 5/2, \pm 10$. Figure 3.11a displays the F_+ and 3.11b the F_- solution sets $\{Z, \sigma\}$. The 'plus' system is simply a nested version of the F_+ result in Fig. 3.10. By contrast, the 'minus' system displays a distinct phase transition with n.

These are, essentially, different possible 'regression model' analogs for cognitive system dynamics under Clauswitzian 'fog' (Wallace 2020).

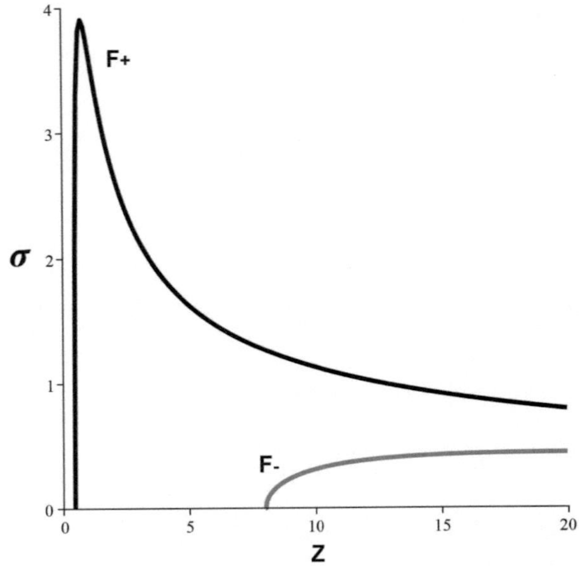

Fig. 3.10 Solution set $\{Z, \sigma\}$ for Fisher Zero phase transitions to 'hallucination' and 'panic' modes under Eq. (3.21), setting $C_1 = -1$, $C_2 = C_3 = 1$. For F_-, the 'hallucination' mode emerges at a very low value of 'noise' σ, while F_+ displays a bifurcated pattern across increasing σ

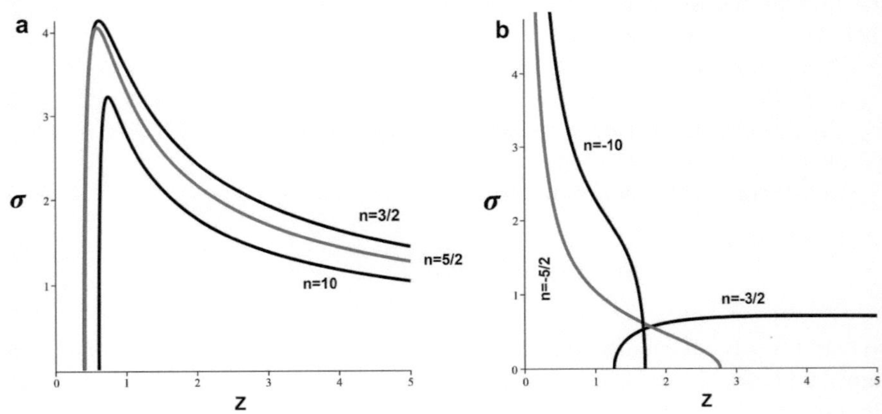

Fig. 3.11 a For positive n, the $\{Z, \sigma\}$ solution sets are simply a nested version of the $F_+ = \exp[-1]$ result of Fig. 3.10. b For negative n, the solution sets show a distinct phase transition with increasing n

For completeness, in Fig. 3.12 we reprise the calculation for Eq. (3.25) as a three-dimensional graph in the solution set $\{\sigma, n, Z\}$, again taking $C_1 = -1$, $C_2 = C_3 = 1$. The complexity of phase transition dynamics under this model is manifest.

It is interesting to carry through this calculation for a different 'fundamental underlying probability distribution' than the Boltzmann model of Eq. (3.22).

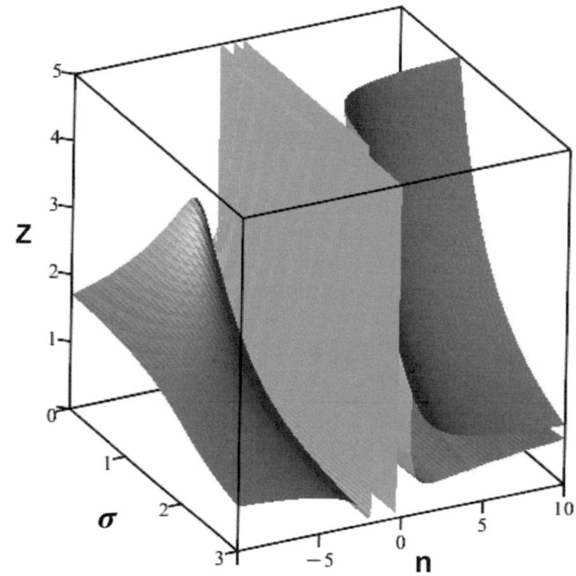

Fig. 3.12 Three-dimensional version of Fig. 3.11, the solution set to Eq. (3.25). Again, $C_1 = -1,\ C_2 = C_3 = 1$

We impose a 'hazard rate'—here, the signal detection rate versus load—as

$$\frac{\rho(x)}{1 - \left(\int_0^x \rho(u)\,\mathrm{d}u\right)} = \frac{1}{1+x}$$
$$\rho(x) = \frac{1}{(1+x)^2} \tag{3.26}$$

noting that the 'hazard rate' for the exponential distribution $m\ \exp[-mx]$ is just m.

Then

$$\frac{1}{(1+F/g)^2} = g$$
$$g_{\pm} = -F + \frac{1}{2} \pm \frac{1}{2}\sqrt{1-4F} \tag{3.27}$$

and Eq. (3.25) becomes

$$F(Z) = \frac{C_1 Z^{-n}}{(n+1)\,n} - \frac{\sigma^2 n\, Z^2}{4n+8} + C_2 Z + C_3 = \frac{1}{4} = 0.25 \tag{3.28}$$

For the Cauchy Distribution

$$\rho(\gamma,\, x) = \frac{2}{\pi}\left(\frac{\gamma}{\gamma^2 + x^2}\right)$$

$$\rho(\gamma, F/g) = g$$
$$g_{\pm} = \frac{1}{\pi\gamma}\left(1 \pm \sqrt{1 - \pi^2 F^2}\right) \tag{3.29}$$

The critical condition then becomes

$$F(Z) = \frac{C_1 Z^{-n}}{(n+1)\,n} - \frac{\sigma^2 n\, Z^2}{4n+8} + C_2 Z + C_3 = \pm\frac{1}{\pi} \approx \pm 0.3183 \tag{3.30}$$

Note that the two latter distributions are 'fat tailed', not having finite first or second moments.

Not unexpectedly, perhaps, for the same values of the C_i, the graphs of Eq. (3.28) and both forms of Eq. (3.30) are recognizably similar to Fig. 3.12.

Stochastic Differential Equations

Einstein's classic analysis of Brownian motion for a large number of particles N leads to the relations

$$\partial\rho(x, t)/\partial t = \mu\partial^2\rho(x, t)/\partial x^2$$
$$\rho(x, t) = \frac{N}{\sqrt{4\pi\mu t}}\exp[-\frac{x^2}{4\mu t}]$$
$$\langle x^2\rangle \propto t \tag{3.31}$$

where t is the time, x is a location variate, and μ a 'diffusion coefficient'.

The last relation represents the average value of x for a particle across the jittering system.

More generally, as a consequence of this insight, it is possible to write, for a Brownian 'stochastic differential' dB_t that might be seen as perturbing some base function, the fundamental relation

$$\langle dB_t^2\rangle = dt \tag{3.32}$$

and on this result hangs much of the subtlety of the approach.

Here, we follow the development of Cyganowski et al. (2002, Sect. 8.4), who provide a series of programs in the computer algebra system MAPLE.

We are given a base 'ordinary' differential equation $dX/dt = f(t, X(t))$ perturbed by a Brownian stochastic variate dB_t according to Eq. (3.33). Tht is, the 'perturbed' solution X_t solves the stochastic differential equation (SDE)

$$dX_t = f(t, X_t)dt + g(t, X_t)dB_t \tag{3.33}$$

We are then given a function Y that depends on both t and X_t. Some calculation—based on Eq. (3.33)—finds that $Y(t, X_t)$ solves the SDE

$$dY_t = L^0 Y_t dt + L^1 Y_t dB_t \tag{3.34}$$

where L^0 and L^1 are the operators

$$\begin{aligned} L^0 Y &= \partial Y/\partial t + f \partial Y/\partial x + \frac{1}{2} g^2 \partial Y^2/\partial x^2 \\ L^1 Y &= g \partial Y/\partial x \end{aligned} \tag{3.35}$$

This is the Ito Chain Rule.

To prove this, as Cyganowski et al. (2002, Sec.8.4) show, one expands

$$\Delta Y_t = Y(t + \Delta t, X_t + \Delta X_t) - Y(t, X_t)$$

to second order, using $\Delta X_t = f \Delta t + g \Delta B_t + \cdots$. Some tedious algebra produces a second-oder term in B_t

$$\begin{aligned} &\frac{1}{2} g^2 \partial^2 Y/\partial x^2 (\Delta B_t)^2 \rightarrow \\ &\frac{1}{2} g^2 \partial^2 Y/\partial x^2 dt \end{aligned} \tag{3.36}$$

by Eq. (3.32) that is then brought into the 'dt' part of the expression. This is one source of the inherent strangeness of stochastic differential equations.

We are centrally concerned with 'nonequilibrium steady states' averaged across the stochastic jitter and represented as

$$\langle dY_t \rangle = L^0 Y = 0 \tag{3.37}$$

A simple MAPLE computer algebra program for the Ito Chain Rule is given in Cyganowski et al. (2002, p. 238), adapted here as

```
L0 := proc(X,a,b) local Lzero, U;
Lzero := diff(U(x,t),t)+a*diff(U(x,t),t,x)+
1/2 * b*b * diff(U(x,t), x, x);
eval(subs(U(x,t)=X,Lzero));
end:
```

where `X` is the expression of Y in Eq. (3.35) in terms of `x`. `a` is $f(t, X_t)$ and `b` is $g(t, X_t)$ in Eq. (3.35), likewise expressed in terms of the base-variable `x`.

References

Atlan H., I. Cohen, 1998, Immune information, self-organization, and meaning, International Immunology, 10:711–717.

Barabasi, A., R. Albert, 1999, Emergence of scaling in random networks, Science 286:509–512.

Bennett, C.H., 1982, The thermodynamics of computation, International Journal of Theoretical Physics 21:905–940.

Butlin, P., et al., 2023, Consciousness in Artificial Intelligence: Insights from the Science of Consciousness, https://arxiv.org/abs/2308.08708.

Cover, T., J. Thomas, 2006, Elements of Information Theory, Second Edition, Wiley, New York.

Cyganowski, S., P. Kloeden, J. Ombach, 2002, From Elementary Probability to Stochastic Differential Equations with MAPLE, Springer, New York.

de Groot, S., P. Mazur, 1984, Nonequilibrium Thermodynamics, Dover, New York.

Derman, E., M. Miller, D. Park, 2016, The Volatility Smile, Wiley, New York.

Dretske, F., 1994, The explanatory role of information, Philosophical Transactions of the Royal Society A 349:59–70.

Effros, M., P. Chou, R. Gray, 1994, Variable-rate source coding theorems for stationary nonergodic sources, IEEE Transactions on Information Theory 40:1920–1925.

Feynman, R., 2000, Lectures on computation,Westview Press, New York.

French, V., E. Anderson, G. Putman, T. Alvager, 1999, The Yerkes-Dodson law simulated with an artificial neural network, Cognitive Systems 5:136–147.

Gould, S.J., 2002, The Structure of Evolutionary Theory, Harvard University Press, Cambridge MA.

Harush, U., B. Barzel, 2017, Dynamic patterns of information flow in complex networks, Nature Communications 8:2181

Jackson, D., A. Kempf, A. Morales, 2017, A robust generalization of the Legendre transform for QFT, Journal of Physics A 50:225201.

Jin, H., Z. Hu, X. Zhou, 2008, A convex stochastic optimization problem arising from portfolio selection, Mathematical Finance 18:171–183.

Khinchin, A., 1957, Mathematical Foundations of Information Theory, Dover, New York.

Laidler, K., 1987, Chemical Kinetics, 3rd ed, Harper and Row, New York.

Landau, L., E. Lifshitz, 2007, Statistical Physics, 3rd Ed. Part 1, Elsevier, New York.

Maturana, H., F. Varela, 1980, Autopoiesis and Cognition: The Realization of the Living, Reidel, Boston.

Nair, G., F. Fagnani, S. Zampieri, R. Evans, 2007, Feedback control under data rate constraints: an overview, Proceedings of the IEEE 95, 108138.

Nocedal, J., S. Wright, 2006, Numerical Optimization, Second Edition, Springer, New York.

Protter, P., 2005, Stochastic Integration and Differential Equations, Second Edition, Springer, New York.

Robinson, S., 1993, Shadow prices for measures of effectiveness II: General model, Operations Research 41:536–548.

Shields, P., D. Neuhoff, L. Davisson, F. Ledrappier, 1978, The Distortion-Rate function for nonergodic sources, The Annals of Probability 6:138–143.

Taleb, N. N. (ed.), 2020, Statistical Consequences of Fat Tails: Real world preasymptotics, epistemology, and applications, STEM Academic Press.

Wallace, R., G. Fricchione, 2024, Stress-induced failure of embodied cognition: A general model, BioSystems, https://doi.org/10.1016/j.biosystems.2024.105193.

Wallace, R., 2012, Consciousness, crosstalk, and the mereological fallacy: an evolutionary perspective, Physics of Life Reviews, 9:426–453.

Wallace, R., 2020, Cognitive Dynamics on Clausewitz Landscapes: The control and directed evolution of organized conflict, Springer, New York.

Wallace, R., 2021, How AI founders on adversarial landscapes of fog and friction, Journal of Defense Modeling and Simulation, https://doi.org/10.1177/1548512920962227.

Wallace, R., (ed.), 2022a, Essays on Strategy and Public Health: The systematic reconfiguration of power relations, Springer, New York.
Wallace, R., 2022b, Consciousness, Cognition and Crosstalk: The evolutionary exaptation of nonergodic groupoid symmetry-breaking, Springer, New York.
Wallace, R., 2023a, Essays on the Extended Evolutionary Synthesis: Formalizations and extensions, Springer, New York.
Wallace, R., 2023b, On 'Machine Consciousness', Journal of Artificial Intelligence and Consciousness, 10:125–148, https://www.worldscientific.com/doi/10.1142/S2705078522500187.
Watts, D., S. Strogatz, 1998, Collective dynamics of small world networks, Nature 393:440–442.

Chapter 4
A Reformulation: Culture Itself as an Information Source

Abstract We reconsider the basic theory of nonergodic cognition, using the homology between nonergodic information source uncertainty and free energy to re-derive peculiar forms of Onsager's nonequilibrium thermodynamics that act as 'regression models' to characterize the dynamics of cognition in 'cultural artifacts'. The result is an extension of the formal perspective beyond the Rate Distortion Control Theory of earlier chapters.

4.1 Introduction

Embodied entities are built from crosstalking cognitive submodules. These range from individual cells, organs, social groupings, formal institutions, to embedding cultures and other environments. For humans in particular, every scale and level of organization, individuals and their social workgroups are constrained, not only by their own experience and training, but by the culture in which they are embedded and with which they interact (e.g., Wallace 2018, 2020a).

They are likewise constrained by the environment in which they operate, including actions and intents of other competing and cooperating entities.

There is, in addition, always structured uncertainty imposed by the large deviations possible within the overall system, including, but not limited to, the embedding environment.

Thus, a number of factors interact to build a composite information source (Cover and Thomas 2006) representing embodied cognition:

- Cognition requires choice that reduces uncertainty and implies the existence of an information source formally 'dual' to that cognition at each scale and level of organization (Atlan and Cohen 1998). The argument is direct and agnostic about representation.
- Cognition requires regulation. As Wallace (2017, Chap. 3) puts it,

> Cognition and its regulation... must be viewed as an interacting gestalt, involving not just an atomized individual, but the individual in a rich context... There can be no cognition without regulation, just as there can be no heartbeat without control of blood pressure,

R. Wallace, *Hallucination and Panic in Autonomous Systems*, Studies in Computational Intelligence 1217, https://doi.org/10.1007/978-3-031-95207-4_4

and no multicellularity without control of rogue cell cancers. Cognitive streams must be constrained within regulatory riverbanks.

It is here that the Data Rate Theorem, or an appropriate generalization, becomes manifest: there must be an embedding regulatory information source imposing control information at a rate greater than an inherently unstable cognitive process generates its own ‘topological information’.

- Most particularly for humans—the colonial naked mole rats of primates—embedding culture is also an information source, with analogs to grammar and syntax. Within any given culture, under particular circumstances, some sequences of behavior are highly probable, and others have vanishingly small probability (Khinchin 1957), a sufficient condition for the development of an equivalence class groupoid symmetry-breaking formalism.
- Spatial and social geographies are similarly structured so as to have incident sequences of very high and very low probability: night follows day, Summer’s dirt roads are followed by October’s impassible mud streams.
- Large deviations, as described by Champagnat et al. (2006) and Dembo and Zeitouni (1998), follow high probability developmental pathways governed by entropy-like laws that imply the existence of another information source.

Embedded and embodied cognition is then characterized by a joint information source uncertainty as (Cover and Thomas 2006)

$$H(\{X_i\}, X_V, X_\Delta) \tag{4.1}$$

The set $\{X_i\}$ includes the cognitive, regulatory, and embedding cultural information sources of the hierarchical system, X_V is the information source of the embedding environment, that may include the actions and intents of adversaries or collaborators, as well as ‘weather’. Finally, X_Δ is the information source of the associated large deviations possible to the system.

Crosstalk between coresident information channels and sources is inevitable, a consequence of the information chain rule (Cover and Thomas 2006). For a set of interacting stationary ergodic information sources X_i, $i = 1, 2, \ldots$, the joint uncertainty of interacting sources and channels is always less than or equal to the sum of the independent uncertainties (Cover and Thomas 2006):

$$H(X_1, X_2, \ldots) \leq \sum_i H(X_i) \tag{4.2}$$

Each information source X_i is powered by some corresponding free energy source $\mathcal{M}_i$, and it takes more free energy to isolate information sources and channels than to allow their interaction. Such a ‘second law’ conundrum confounds much of electrical engineering, particularly in the design and construction of microchips. Such ‘second law’ problems extend to all scales and levels of organization.

Evolutionary process has taken this 'spandrel', in the sense of Gould and Lewontin (1979), and built new cathedrals from it (Wallace 2012).

The next steps are not entirely straightforward.

According to popular mathematical canon, there is really no serious work to be done on nonergodic information sources as a consequence of the Ergodic Decomposition Theorem (e.g., Gray 1988, Chap. 7) which states that it is possible to factor any nonergodic process into a sufficiently large sum (or generalized integral) of ergodic processes, in the same way that any point on a triangle can be expressed in terms of its extremal fixed point vertexes. As Winkelbauer (1970) put it for information source uncertainty,

Theorem II *The asymptotic rate of a stationary source* μ *equals the essential supremum of the entropy rates of its ergodic components:*

$$H(\mu) = ess. \sup_{z \in R[\mu]} H(\mu_z)$$

where the μ_z *are ergodic.*

Is this really a 'simple' result for dynamic systems that can suffer 'absorbing states'? Individual paths—and small, closely-related, equivalence classes of them—are particularly important in biological phenomena, as opposed to physical process. This is because each path may have a unique consequence for the organism or other entity embedded in a stressful environment. There will, for example, only be a single 'meaningful sequence' associated with successful capture by a predator: absorbing states are particularly important in biological processes.

Recall that it is possible to approximate any reasonably well-behaved real-valued function over a fixed interval in terms of a Fourier series. It was, in the geocentric Ptolemaic system, via a sufficient number of epicycles, possible to predict planetary positions to any desired accuracy using such a de-facto Fourier Decomposition. The underlying astronomical problem was both simplified and enhanced by the non-geocentric empirical observations of Kepler, explained by Newton, and fully elaborated by Einstein.

Cognition is considerably more complex than the motion of the planets around the sun, and Keplerian laws must still be found across many different physiological phenomena and organisms. Newton and Einstein are nowhere on the horizon.

4.2 Reformulating the Theory

Next, we expand the development of Wallace (2018), deriving from first principles—rather than imposing—a 'temperature' measure for nonergodic cognitive systems.

We have, in previous chapters, reduced the spectrum of resources and their interactions—including internal bandwidth, rates of sensory information, and material/energy supply—in terms of a scalar rate variable Z.

To explore the dynamics of cognition, we again introduce a first-order linear Onsager approximation abducted from nonequilibrium thermodynamics (de Groot and Mazur 1984).

Again, we invoke an iterated free energy Morse Function (Pettini 2007) via a formalized Boltzmann probability expression, in the sense of Feynman (2000). This is done by enumerating high probability developmental pathways available to the system as $j = 1, 2, \ldots$, allowing definition of a path probability P_j

$$P_j = \frac{\exp[-H_j/g(Z)]}{\sum_k \exp[-H_k/g(Z)]} \tag{4.3}$$

As described above, other probability distributions can be assumed, at the cost of computational complexities. Here, we fixate on the Boltzmann for convenience.

This formulation, following Khinchin (1957) and Wallace (2018), will apply to stationary nonergodic as well as to stationary ergodic information sources and can be used for systems in which each developmental pathway x_j has its own source uncertainty measure H_{x_j}. This, however, is only expressed as a Shannon 'entropy' for an ergodic system (Khinchin 1957).

Note that Eq. (4.3) differs significantly from the Rate Distortion Control Theory formulation of previous chapters: H, as a joint information source, is not R, a control channel capacity associated with a particular form of average distortion D.

We are, in some measure, liberated from the kind of dilemma Tegmark (2016) explores for Integrated Information Theory in that we do not need the functional form of a H_k across a probability distribution, only it's scalar magnitude. Further development, however, involves imposition of an 'adiabatic approximation' that assumes a 'sufficiently stationary' convergence about slowly-varying underlying system dynamics, much as is done in the study of molecular quantum mechanics, where rapidly-changing electron dynamics are approximated around slowly-changing nuclear configurations. Previous work—Wallace (2005, 2012)—specified 'adiabatically, piecewise, stationary ergodic' (APSE) systems that remained as close to stationary and ergodic as needed for the asymptotic limit theorems to work. Here, we lift the ergodic requirement.

The 'temperature' $g(Z)$, like the loss of ergodicity, also becomes another matter, and must now be calculated from Onsager-like system dynamics built from the partition function, i.e., from the denominator of Eq. (4.3). This implies its own set of approximations.

The system's 'rate of cognition' can then be expressed, as in chemical reaction theory (Laidler 1987), by the probability such that $H_j > H_0$, where H_0 is the lower limit for detection of a signal under embodiment in a varying and noisy environment, or for stability via the Data Rate Theorem.

There is an implicit point: The 'prime groupoid phase transition'. We are, in essence, imposing a symmetry-breaking version of the Source Coding Theorem, the Shannon-McMillan Theorem, dividing all possible paths into a small set—an equivalence class—of 'meaningful' sequences consonant with some underlying grammar and syntax—in a large sense—and a very large set of vanishingly small probability paths that are not consonant.

Such an equivalence class partition imposes the first of many groupoid symmetry breaking phase transitions. Groupoid symmetries are discussed in more detail in the Mathematical Appendix, but, in essence, we have used a groupoid symmetry-breaking phase change to derive—or to impose—a fundamental information theory asymptotic limit theorem. Other such may emerge 'naturally' from related groupoid symmetry-breaking phenomena, all related to equivalence classes of system developmental pathways.

This result can be seen, from a biological perspective, as in the same ballpark as a much earlier phase transition, the sudden transmission of light across the primordial universe after the first 370,000 years.

In sum, groupoid symmetry-breaking extends the Shannon-McMillan Source Coding Theorem to nonergodic—and possibly non-stationary—information sources. This is a major—if 'trivially obvious'—result to which we will return below. Again, the Mathematical Appendix provides a brief introduction to the standard groupoid algebra (Brown 1992; Cayron 2006; Weinstein 1996). The central matter is that products are not defined for all possible pairs of elements, leading to disjoint orbit partition.

The iterated free energy Morse Function F is—again—defined by the relation

$$\exp[-F/g(Z)] \equiv \sum_k \exp[-H_k/g(Z)] = h(g(Z)) \tag{4.4}$$

again fixating on the Boltzmann distribution for convenience.

F is a Morse Function subject to symmetry-breaking transitions as $g(Z)$ varies (Pettini 2007).

We reiterate that these symmetries are not those associated with simple physical phase transitions represented by standard group structures. Cognitive phase change involves punctuated transitions between equivalence classes of high probability signal sequences, represented as groupoids. As Tateishi et al. (2013) put it, if experimental data can be grouped into equivalence classes compatible with an algebraic structure, a groupoid approach can capture the symmetries of the system in a way not be possible with group theory, for example in the analysis of neural network dynamics. Deeper delvings into similar matters can be found in Schreiber and Skoda (2010).

Here, groupoid symmetries are driven by the directed homotopy induced by failure of local time reversibility for information systems. This is because palindromes have vanishingly small probability. In English, 'the' has meaning in context while 'eht' has vanishingly low probability. The ubiquitous mathematical trope 'except on a set of measure zero' implies a fundamental symmetry breaking.

Particularly complicated cognitive systems may require even more general structures, for example, small categories and/or semigroupoids for analogs to the standard symmetry-breaking dynamics of physical systems.

These more general symmetry-breaking phase changes represent extension of the Data Rate Theorem (DRT) to cognitive systems. Again, the DRT states that the rate at which externally-supplied control information must be imposed on an inherently unstable system to stabilize it must exceed the rate at which that system generates its own 'topological information'. The model is of steering a vehicle on a rough, twisting roadway at night. The headlight/steering/driver complex must impose control information at a rate greater than the 'twistiness/roughness' of the road imposes it's own information on the vehicle.

There may be many 'phases' available to a cognitive system as $g(Z)$ varies, rather than just the 'on/off' of stability implied by the DRT itself.

Dynamic equations can be derived from Eq. (4.3) by invoking the usual first-order Onsager approximation in the gradient of an entropy measure constructed from the iterated free energy Morse Function F via the Legendre transform

$$S(Z) \equiv -F(Z) + Z\mathrm{d}F/\mathrm{d}Z \tag{4.5}$$

This leads to the relations

$$\begin{aligned}
\exp[-F/g(Z)] &= \sum_k \exp[-H_k/g(Z)] \equiv h(g(Z)) \\
F(Z) &= -\log(h(g(Z))g(Z) \\
g(Z) &= -\frac{F(Z)}{RootOf\left(\mathrm{e}^X - h(-F(Z)/X)\right)} \\
\partial Z/\partial t &\approx \mu \partial S/\partial Z = f(Z)
\end{aligned} \tag{4.6}$$

where the *RootOf* construct defines a generalized Lambert W-function in the sense of Maignan and Scott (2016); Mezo and Keady (2015), and Scott et al. (2006).

The last expression in Eq. (4.6) represents imposition of the entropy gradient formalism of Onsager nonequilibrium thermodynamics (de Groot and Mazur 1984), for which $f(Z)$ is the 'adaptation function', the fundamental rate at which the system adjusts to changes in Z. Again, for information transmission there is no 'temporal microreversibility', so that there can be no Onsager Reciprocal Relations in these models.

We, somewhat arbitrarily, set the 'diffusion coefficient' μ equal to 1.

After further development,

$$\begin{aligned}
&f(Z) = Z\mathrm{d}^2F/\mathrm{d}Z^2 \\
&F(Z) = \int\int \frac{f(Z)}{Z}\mathrm{d}Z\mathrm{d}Z + C_1 Z + C_2 \\
&-Z\int \frac{f(Z)}{Z}\mathrm{d}Z - \log(h(g(Z)))g(Z) - C_1 Z + \int (f(Z)\mathrm{d}Z + C_2 = 0
\end{aligned} \tag{4.7}$$

Again, taking $F = -\log(h(g(Z))g(Z)$, with h determined by underlying internal structure, leads to expressing the 'temperature' g in terms of a generalized Lambert W-function, suggesting an underlying formal network structure (Newman 2010).

Specification of any two of f, g, h allows calculation of the third. Note, however, that h is fixed by the internal structure of the larger system, and the 'adaptation rate' f is imposed by externalities. In addition, the 'boundary conditions' C_1, C_2 are likewise externally-imposed, also structuring the temperature-analog $g(Z)$. Indeed, the 'temperature' $g(Z)$ might well be viewed as itself an order parameter.

We assume here that embodied cognitive systems can be characterized by the scalar parameter Z, mixing material resource/energy supply with internal and external flows of information under time constraint. There may be more than one such composite irreducible entity driving system dynamics. More explicitly, it may be necessary to replace the scalar Z with some $m \leq n$-dimensional vector having a number of independent—even orthogonal—components accounting for considerable portions of the total variance in the rate of supply of essential resources. The dynamic equations can then be reexpressed in a more complicated vector form.

In a similar way, it may be necessary to introduce nonlinear or higher order Onsager models. An introduction to these matters can be found in Wallace (2021) and in Chap. 3 involving, for example, expressions of the form

$$
\begin{aligned}
S &= -F + \sum_j a^j Z^j \mathrm{d}^j F/\mathrm{d}Z^j \\
\partial Z/\partial t &\approx \sum_j b_j d^j S/\mathrm{d}Z^j = f(Z) \\
\partial Z/\partial t &\approx \mathrm{d}S/\mathrm{d}Z \times d^2 S/\mathrm{d}Z^2 + \cdots = f(Z)
\end{aligned}
\tag{4.8}
$$

leading to formal algebraic power series treatments in the sense of Jackson et al. (2017). These considerations take us beyond the simplest '$Y = mX + b$' regression model analogs.

The dynamics are driven at rates determined by the adaptation function $f(Z)$. We can ask more detailed questions regarding what happens at critical points defined in terms of the 'temperature' variate $g(Z)$ through the abduction of another approach from physical theory.

4.3 Further Development

We have—implicitly—focused here largely on rapid, single-workspace phenomena analogous to neural consciousness in higher animals, having the 100ms time constant. If we enthrone the Data Rate Theorem rather than rate of cognition, we can say something regarding embodied cognition as a—or even the—more fundamental gestalt. The argument—a significant condensation of Wallace (2021)—is surprisingly simple, and seems independent of such niceties as ergodicity of information sources.

Recall the Data Rate Theorem. Stabilization of an inherently unstable control system engaged in some fundamental task requires that control information be delivered at a rate greater than the rate at which the unstable system generates its own 'topological information', say H_0. We do not specify 'H_0', except as a scalar entity that may indeed change with time.

Most simply, we characterize H as the (scalar) rate at which external regulatory mechanisms provide such control information, and (yet again) define a Boltzmann pseudoprobability

$$dP(H) \equiv \frac{\exp[-H/g(Z)]dH}{\int_0^\infty \exp[-H/g(Z)]dH} \tag{4.9}$$

where $g(Z)$ and Z are as described above, supposing that the regulatory-action process is itself composed of subcomponents that interact with each other and with an embedding environment through both information exchanges and the use of 'materiel' in various forms.

We define iterated free energy and entropy analogs as above from the 'partition function' denominator of Eq. (4.9), obtaining the usual relations

$$\begin{aligned}
\exp[-F/g(Z)] &\equiv \int_0^\infty \exp[-H/g(Z)]dH = g(Z) \\
g(Z) &= \frac{-F(Z)}{W(n, -F(Z))} \\
S &\equiv -F + ZdF/dZ \\
\partial Z/\partial t &\propto dS/dZ = f(Z) \\
f(Z) &= Zd^2F/dZ^2 \\
L(Z) &= \frac{\int_{H_0}^\infty \exp[-H/g(Z)]dH}{\int_0^\infty \exp[-H/g(Z)]dH} = \exp[-H_0/g(Z)]
\end{aligned} \tag{4.10}$$

where, again, $W(n, x)$ is the appropriate Lambert W-function and $L(Z)$ the cognition rate. A more convoluted line of argument leads to the 'generalized Lambert W-functions' of Eq. (4.6).

Embodiment—directly implying interaction with the embedding world—appears to impose a certain draconian or Procrustean simplicity, as Charles Darwin, Alfred Russel Wallace, and many others have noted.

4.4 Regulation and Hallucination

A somewhat convoluted argument recovers the cognition/regulation dyad of the previous chapter in terms of the 'cultural' perspective. This requires examination of stochastic dynamics under the assumption of Eq. (4.10) that $dZ/dt = Zd^2F/dZ^2$, where F is the free energy analog.

We take the system as driven by the stochastic differential equation

$$dZ_t = \left(Z_t d^2F/dZ^2\right) dt + \sigma Z_t dB_t \tag{4.11}$$

where the second term represents volatility and dB_t is Brownian white noise. Then, imposing second-order stability under the Ito Chain Rule, so that $\langle dZ_t^2 \rangle = 0$, straightforward calculation finds

$$F(Z) = -\frac{\sigma^2 Z^2}{4} + C_1 Z + C_2 \tag{4.12}$$

where the C_i are (important) boundary conditions.

We assume that two 'cultural information sources' operate simultaneously, that of the original cognitive system of interest, with source uncertainty H and a regulatory agent that interacts with it, of source uncertainty aH. The partition function expression for F and the cognitive temperature analog g is then

$$\exp[-F(Z)/g] = \exp[-H/g] + \exp[-aH/g] \tag{4.13}$$

This relation can be solved numerically for g in terms of the parameter and variable values, taking F as in Eq. (4.12). For Fig. 4.1, $H = 1$, $C_1 = -1$, $C_2 = 1$ while σ and Z vary. $a = 0, 2$ as indicated.

For $a = 0$, the region to the left of the surface represents complex-valued 'hallucination' and/or other instabilities. Sufficient regulation a, however, eliminates such instabilities in this model.

4.5 Multisignal Overload

In contrast to the previous section, for which (i), stability in variance is assumed through Eq. (4.12), and (ii), the probability model is described by Eq. (4.13), we explore a system overwhelmed by multisignal overload. That is, cognition's underlying probability model is strongly dominated by a plethora of independent and competing information sources H_i, $i = 1, \ldots$, so that the probability associated with an individual source j becomes

$$P_j = \frac{H_j}{\sum_i H_i} = \frac{H_j}{H} \tag{4.14}$$

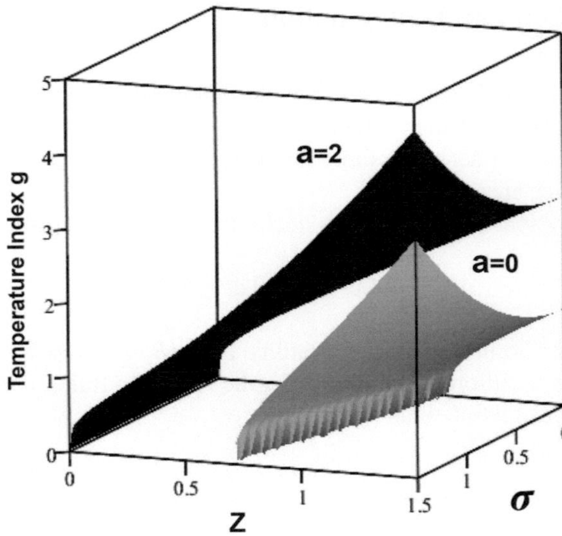

Fig. 4.1 Numerical solution to Eq. (4.13) for the temperature analog g, taking $H = 1$, $C_1 = -1$, $C_2 = 1$. Z and σ vary. For $a = 0$, the region to the left of the surface represents complex-valued 'hallucination' or other forms of instability. Sufficient regulation a, however, eliminates such instabilities in this model

where the H_i are the individual information source uncertainties.

$H \equiv \sum_i H_i$ then becomes the basic partition function driving dynamics.

Adopting a simple Boltzmann model for free energy F,

$$\begin{aligned} \exp[-F/g] &= H \\ g &= \frac{-F}{\ln(H)} \\ L &= H^{H_0/F} \end{aligned} \tag{4.15}$$

where the third form follows from the last expression Eq. (4.10), and H_0 is the system trigger threshold.

Like all versions of the exponential distribution, the Boltzmann Distribution has a fixed 'hazard rate' at which, here, signals are detected as a function of signal strength. In the chapter Mathematical Appendix we explore the complexities arising in systems for which the hazard rate declines or increases monotonically with rising signal.

Circumstance now does not guarantee or impose second-order stability, and, again following Eq. (4.10), we take $dZ/dt = Fd^2F/dZ^2 = f(Z) = \beta - \alpha Z$, the 'exponential model'. Then

$$F(Z) = \beta(Z\ln(Z) - Z) - \frac{\alpha Z^2}{2} + C_1 Z + C_2 \tag{4.16}$$

Figure 4.2a, shows cognition rate as a function of 'arousal' β, with $\alpha = H_0 = 1$, $C_1 = -1$, $C_2 = 3$. We again obtain a the canonical inverse-U Yerkes-Dodson relation of cognition rate with arousal, book-ended by 'hallucination' on the left and 'panic' on the right. Here, however, the phase transitions are very highly, indeed, pathologically, punctuated.

Even for the peak stable condition, where $\beta = 3$, there is a serious problem. Figure 4.2b shows that the maximum stable cognition rate collapses at sufficient H, essentially a measure of the degree of confusion brought to cognitive dynamics by simultaneous impingement of a burdensome number of independent and uncoordinated information sources. These become a form of obfuscating noise rather than sources of useful intelligence. One is again reminded of 'Maskirovka' (e.g., Wallace 2021b, 2022).

It is of interest to explore possible stochastic extensions in which the cognition rate in Eq. (4.15) is assumed to be in a nonequilibrium steady state (nss). System dynamics are then taken as driven by, respectively, the two *independent* stochastic differential equations

$$\begin{aligned} dZ_t &= (\beta - \alpha Z_t)dt + \sigma Z_t dB_t^1 \\ dH_t &= (b - a H_t)\, dt + \omega H_t dB_t^2 \end{aligned} \tag{4.17}$$

where, assuming ordinary volatility for the second terms, the dB_t^j are standard Brownian white noise (Protter 2006). Figure 4.2c, using the third of Eq. (4.15), takes parameter values as in a and b, with $\beta = 3$, and applies the Ito Chain Rule to impose the nss condition $\langle dL_t \rangle = 0$. The necessary condition solution set $\{\sigma, H, Z\}$ is shown. Values outside this solution set are unstable. Note that rising 'noise' σ imposes progressive difficulty in parsing the sensory/internal information rates represented by Z.

Similarly, Fig. 4.2d carries forward the calculation for $\langle dL_t \rangle = 0$ for the second form of Eq. (4.17), where H itself is subject to stochastic variation, while Z is fixed. Rising ω drives down the level of toleration for multisignal overload, as measured by H.

Section 4.3 was based on a probability model of the form $\rho(F/g) = g$, while, in this section, $\rho(F/g) = H$, where $\rho(x)$ is the underlying fundamental probability distribution, here taken as the Boltzmann. The chapter Mathematical Appendix first explores a more general situation in which $\rho(F/g) = V(g, H)$, suggesting that the probability models developed here can be used to construct statistical tools for application across a broad spectrum of experimental and observational schema.

A final model extension in the chapter Appendix explores two simple *joint* stochastic variations in Z and H according to the relations of Eq. (4.17), under the circumstance that Clausewitzian fog and friction ensure $\sigma = \omega$.

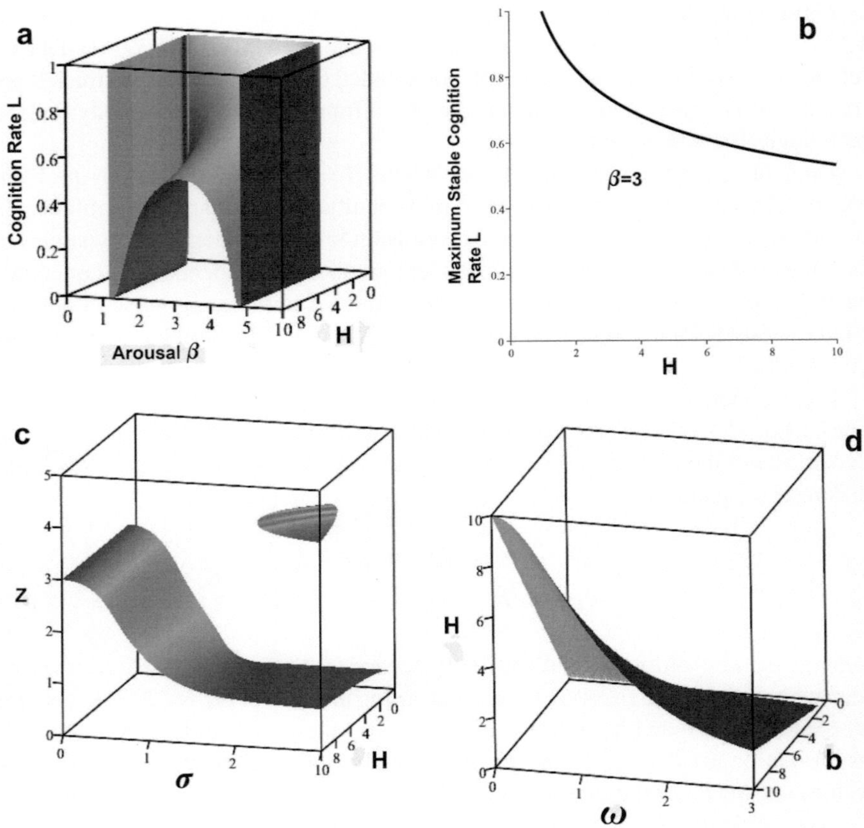

Fig. 4.2 **a** Cognition rate as a function of 'arousal' β. $\alpha = H_0 = 1,\ C_1 = -1,\ C_2 = 3$. Typically, there is an inverse-U Yerkes-Dodson relation of cognition rate with arousal, book-ended by 'hallucination' on the left and 'panic' on the right. The phase transitions, however, are pathologically, punctuated. **b** The maximum stable cognition rate collapses at sufficient H, measuring the degree of confusion brought by simultaneous impingement of too many independent and uncoordinated information sources. **c** Using the third of Eq. (4.15), taking parameter values as in a and b, the Ito Chain Rule imposes the nss condition $\langle dL_t \rangle = 0$ for $\beta = 3$. Parameter values outside this solution set are unstable. Rising 'noise' σ causes progressive difficulty parsing the sensory/internal information rates represented by Z. **d** A similar calculation of the nonequilibrium steady state condition $\langle dL_t \rangle = 0$ for stochastic variation in H, with Z fixed. Rising ω drives down the tolerated level of multisignal overload H

4.6 Recovering the Earlier Formalism

Another characterization is via a Data Rate Theorem (DRT) implementation of the Rate Distortion Theorem (RDT) for non-ergodic information sources. The DRT determines how a control signal u_i is expressed in the system response at time $t + 1$ as x_{t+1}. Adopting the RDT perspective, one deterministically retranslates an observed set of outputs $X^i = \{x_1^i, x_2^i, \ldots\}$ into a sequence of possible control signals $\hat{U}^i = \{\hat{u}_1^i, \hat{u}_2^i, \ldots\}$ and compares the inferred control sequence with the original control sequence $U^i = \{u_1^i, u_2^i, \ldots\}$. A 'difference' between the U is scalarized under the 'distortion measure' H, the control information rate necessary to impose stability on the inherently unstable control system. It is then possible to define an equivalent 'average distortion' $\mathcal{H}$ as

$$\mathcal{H} = \sum_i p(U^i) H(U^i, \hat{U}^i) \tag{4.18}$$

where $p(U^i)$ is the probability of the sequence u^i. $H(U^i, \hat{U}^i)$, the external control information rate needed for stability, is to be taken as the 'information distortion' between the two control sequences.

It is possible to define a convex Rate Distortion Function $R(\mathcal{H})$ for nonergodic information sources in terms of an average across the ergodic components of that source (e.g., Shields et al. 1978; Leon-Garcia et al. 1979; Efros et al. 1994, etc.). This allows construction of another (here, Boltzmann) pseudoprobability in the resource rate Z as

$$dP(R, Z) = \frac{\exp[-R/g(Z)]dR}{\int_0^\infty \exp[-R/g(Z)]dR} \tag{4.19}$$

Again, it is possible to define a free energy analog in terms of the partition function, giving

$$\begin{aligned}
\exp[-F/g(Z)] &= \int_0^\infty \exp[-R/g(Z)]dR = g(Z) \\
F(Z) &= -g(Z)\log[g(Z)] \\
g(Z) &= \frac{-F(Z)}{W(n, -F(Z))}
\end{aligned} \tag{4.20}$$

leading towards another version of the previous chapters' results. Again, $W(n, x)$ is the Lambert W-function of order n.

4.7 Discussion

We have reconsidered the dynamics of the earlier chapters via a joint information source convoluting both information streams dual to cognitive processes and those associated with embedding 'environmental' phenomena including culture and other cultural artifacts. For human, institutional, and machine cognition, the embedding culture and its path-dependent historical trajectory contribute to the form of the 'riverbanks' between/among which the 'stream of cognition' can meander. Those 'riverbanks', of course, are also formed by the nature of the environment embedding the cognitive agents of interest, and also by the acts of other cognitive agencies against which the systems-of-interest may contend, or with whom they cooperate. Indeed, much of what one studies in these matters takes place on Clausewitz landscapes of fog, friction, and deadly adversarial intent (Wallace 2020b).

While we began with information sources, by Eq. (4.20) we have returned to a form of Rate Distortion Control Theory, albeit with an average distortion measured by the control information necessary to impose stability on the composite cognitive system of interest.

Interspersed throughout this new formalism are groupoid symmetry-breaking phase transitions analogous to phase change in physical systems that instantiate, in a different way, the Fisher Zero approach of the earlier chapters. Some of this is studied at greater length in Wallace (2023, Chap. 1).

There are, then, even from the perspective of the asymptotic limit theorems of information and control theories, many ways to climb the scientific trees that confront and confound the study of real-world, real-time embodied cognition and its many pathologies.

Mathematical Appendix

Groupoids

We following Brown (1992). Consider a directed line segment in one component, written as the source on the left and the target on the right.

$$\bullet \longrightarrow \bullet$$

Two such arrows can be composed to give a product **ab** if and only if the target of **a** is the same as the source of **b**

$$\bullet \xrightarrow{\mathbf{a}} \bullet \xrightarrow{\mathbf{b}} \bullet$$

Brown puts it this way,

> One imposes the geometrically obvious notions of associativity, left and right identities, and inverses. Thus a groupoid is often thought of as a group with many identities, and the reason why this is possible is that the product **ab** is not always defined.
>
> We now know that this apparently anodyne relaxation of the rules has profound consequences… [since] the algebraic structure of product is here linked to a geometric structure, namely that of arrows with source and target, which mathematicians call a *directed graph.*

Cayron (2006) elaborates this:

> A group defines a structure of actions without explicitly presenting the objects on which these actions are applied. Indeed, the actions of the group G applied to the identity element

e implicitly define the objects of the set G by ge = g; in other terms, in a group, actions and objects are two isomorphic entities. A groupoid enlarges the notion of group by explicitly introducing, in addition to the actions, the objects on which the actions are applied. By this approach, many identities may exist (they correspond to the actions that leave an object invariant).

Stewart (2017) describes something of the underlying mechanics by which symmetry changes in general may be expressed:

> Spontaneous symmetry-breaking is a common mechanism for pattern formation in many areas of science. It occurs in a symmetric dynamical system when a solution of the equations has a smaller symmetry group than the equations themselves... This typically happens when a fully symmetric solution becomes unstable and branches with less symmetry bifurcate.

It is of particular importance that equivalence class decompositions permit construction of groupoids in a highly natural manner.

Weinstein (1996) and Golubitsky and Stewart (2006) provide more details on groupoids and on the relation between groupoids and bifurcations.

An essential point is that, since there are no necessary products between groupoid elements, 'orbits', in the usual sense, disjointly partition groupoids into 'transitive' subcomponents.

Application of gropoid formalism to the homochirality problem, with a more detailed overview of 'biological' groupoid algebra, can be found in Wallace (2011).

A Declining Hazard Rate Model

A probability distribution $\rho(x)$ has a 'hazard rate' $Q(x)$ defined as

$$Q(x) = \frac{\rho(x)}{1 - \int_0^x \rho(u)du} \tag{4.21}$$

For the Exponential Distribution $m \exp[-mx]$ the hazard rate is just m. That is, 'signal detection', in our sense, is independent of the 'signal load' x. We next study the case for which $Q(x) = 1/(1 + x)$. Solution of the resulting integral equation finds $\rho(x) = 1/(1 + x)^2$, a 'fat-tailed' distribution without finite mean or variance. Then, following Eq. (4.15),

$$\begin{aligned} \frac{1}{(1 + F/g)^2} &= H \\ g_\pm &= \frac{-(H \pm \sqrt{H})F}{H - 1} \\ L &= \frac{g}{g + H_0} \end{aligned} \tag{4.22}$$

Note the two possible forms for the temperature analog g. This generates two different dynamic modes analogous to Fig. 4.2, as shown in Fig. 4.3a and c, similar to Fig. 4.2a. Here, again, from Eq. (4.16), $\alpha = 1,\ C_1 = -1,\ C_2 = 3$ and β varies as the index of 'arousal'.

Imposition of Eq. (4.17), and then calculating the nonequilibrium steady state $\langle dL_t \rangle = 0$ at $\beta = 3$, leads to Fig. 4.3b and d. Compared to Fig. 4.2c, these are both multi-modal, suggesting the possibility of further bifurcation instabilities with rising 'noise'.

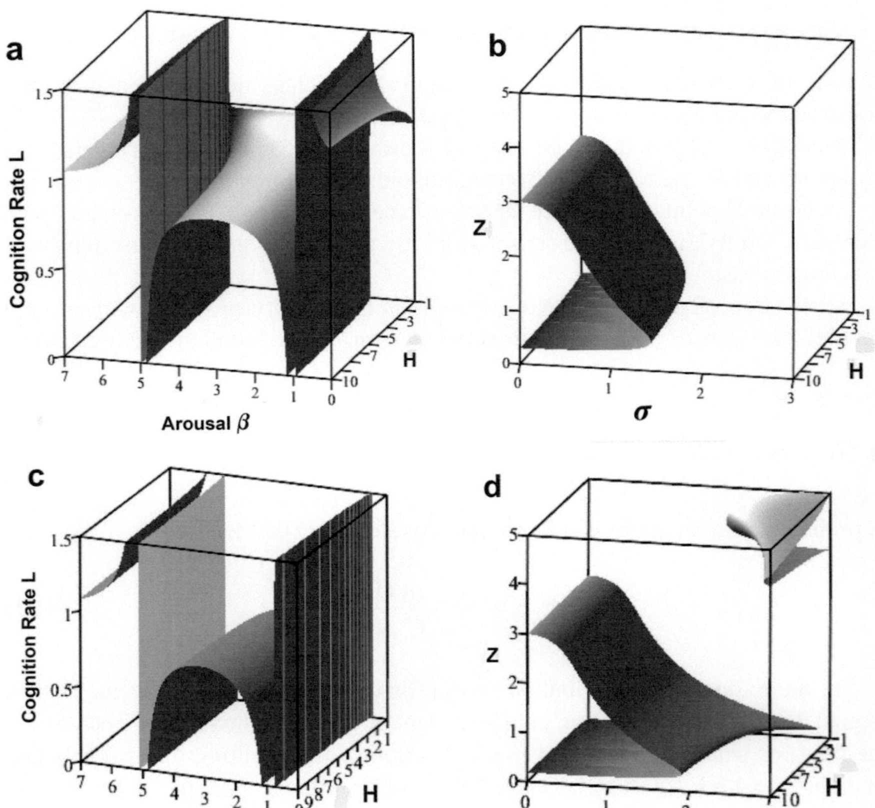

Fig. 4.3 Analogs to Fig. 4.2 for the two different modes available to the distribution $\rho(x) = 1/(1 + x)^2$. Figures **a** and **b** characterize the 'minus' mode. Unlike the example of Fig. 4.2c, this system can undergo further complicated bifurcation instabilities with rising 'noise'

Increasing Hazard Rate Models

For the Rayleigh Distribution,

$$\begin{aligned} \rho(m, x) &= \frac{x}{m^2}\exp[-x^2/2m^2] \\ Q(m, x) &= \frac{x}{m^2} \end{aligned} \tag{4.23}$$

Recall that this distribution can be taken as representing the length of a two-dimensional vector whose components obey the same Normal Distribution.

Setting $\rho(m, F/g) = H$ gives

$$\begin{aligned} g &= \frac{F\sqrt{-\frac{H^2m^2}{W(n,-H^2m^2)}}}{H\,m^2} \\ L &= \exp[-H_0^2/(2g^2m^2)] \end{aligned} \tag{4.24}$$

where $W(n, X)$ is the Lambert W-function of order n, real-valued only for $n = 0, -1$ and $X > -\exp[-1]$. Some algebra shows that a necessary condition for real-valued cognition rate L is $H^2m^2 \leq \exp[-1]$. Within that—singularly narrow—range, the Yerkes-Dodson and stochastic stability patterns are roughly similar to Figs. 4.2 and 4.3.

For the sum of three independent, Normally Distributed vector components, so that $x^2 = x_1^2 + x_2^2 + x_3^2$, one obtains the classic Maxwell-Boltzmann Distribution of ideal gas theory,

$$\begin{aligned} \rho(m, x) &= \sqrt{2/\pi}\frac{x^2}{m^2}\exp[-x^2/(2m^2)] \\ Q(m, x) &\propto x/m^2, \quad x \to \infty \end{aligned} \tag{4.25}$$

Solution to the relation $\rho(m, F/g) = H$ leads to

$$\begin{aligned} g &= -\frac{F}{\sqrt{-2\,W\left(-\frac{H\sqrt{\pi}\sqrt{2}m}{4}\right)}\,m} \\ L &= -\frac{\left(-2\,\mathrm{e}^{-\frac{1}{2g^2m^2}} + \sqrt{\pi}\,mg\sqrt{2}\left(\mathrm{erf}\left(\frac{\sqrt{2}}{2\,gm}\right) - 1\right)\right)\sqrt{2}}{2\sqrt{\pi}\,gm} \end{aligned} \tag{4.26}$$

As for the Rayleigh Distribution, the Lambert W-function introduces a severe constraint on H:

$$Hm\sqrt{2\pi}/4 < \exp[-1] \tag{4.27}$$

Cognition rate and stochastic patterns are similar to those of the Rayleigh Distribution.

Terra Incognita

If, by contrast, one imposes the structure for Z as the 'simple' *product* of three independent normally-distributed components having the same means and variances, then

$$\rho(m, x) = \frac{1}{2\sqrt{2}\pi^{3/2}m^3} G_{0,3}^{3,0}\left(\frac{x^2}{8m^6}|0, 0, 0\right) \tag{4.28}$$

where G is the Meijer-G function of appropriate order. This takes us into very difficult mathematical realms.

In any event, distribution form clearly counts for cognition rate stability under increasing total information load $H = \sum_i H_i$.

Extending the Perspective

Sections 4.3 and 4.5 studied models for which $\rho(F/g) = g$ and $\rho(F/g) = H$, respectively, where $\rho(x) = \exp[-x]$. A mixed situation, where both models apply, might involve $\rho(F/g) = V(g, H)$. Here, we again set $\rho(x) = \exp[-x]$, and let V take the respective forms gH and $g + H$, with F given as in Eq. (4.16) and cognition rate L as in Eq. (4.10). Parameter values as in Fig. 4.2, i.e., $\alpha = H_0 = 1$, $C_1 = -1$, $C_2 = 3$. β becomes the index of 'arousal'.

For the first case, letting $V = gH$,

$$g = \frac{-F}{W(n, -HF)} \tag{4.29}$$

where $W(n, x)$ is again the Lambert W-function—here of order 0—generating the familiar 'simple' Yerkes-Dodson result in Fig. 4.4a.

Taking $V = g + H$, on the other hand, leads to an intractable algebraic relation that must be addressed numerically,

$$F = -\ln(g + H)g \tag{4.30}$$

leading to Fig. 4.4b, where the solution bifurcates into two distinctly different modes, one a familiar Y-D form, and the other instantiating a complex instability.

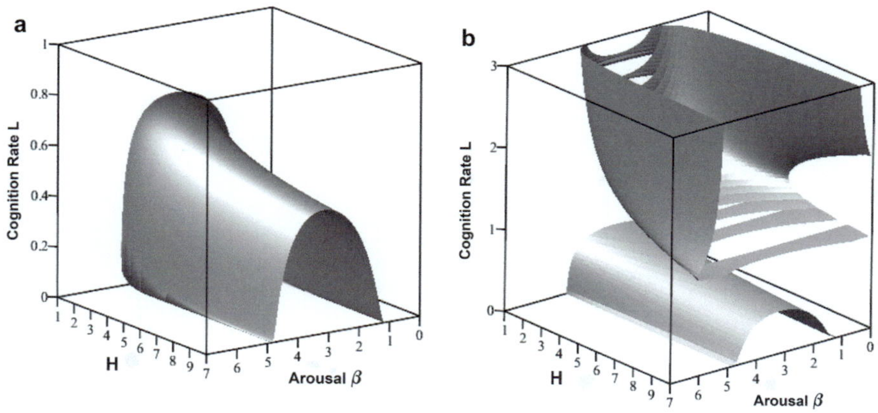

Fig. 4.4 F is as in Fig. 4.2. **a** The relation $V = gH$ is directly solvable in terms of the Lambert W-function, providing the familiar Y-D pattern. **b** $V = g + H$ generates a bifurcated solution set, one the expected Y-D result, but another representing onset of a highly complicated instability

Joint Stochastic Variation

It is tedious but 'not difficult'—using contemporary financial engineering computer algebra packages—to explore the nonequilibrium steady state $\langle dL_t \rangle = 0$ from Eq. (4.15) via application of the Ito Chain Rule to a two-dimensional system based on the simultaneous expression of both models of Eq. (4.17) with and without crosstalk.

We study two examples. The first sets $\omega = \sigma$, includes a classic volatility term, but omits crosstalk:

$$d\begin{pmatrix} H_t \\ Z_t \end{pmatrix} = \begin{pmatrix} b - aH_t \\ \beta - \alpha Z_t \end{pmatrix} dt + \begin{pmatrix} \sigma H_t \\ 0 \end{pmatrix} dB_t^1 + \begin{pmatrix} 0 \\ \sigma Z_t \end{pmatrix} dB_t^2 \tag{4.31}$$

The second invokes simple crosstalk, but omits standard volatility:

$$d\begin{pmatrix} H_t \\ Z_t \end{pmatrix} = \begin{pmatrix} b - aH_t \\ \beta - \alpha Z_t \end{pmatrix} dt + \begin{pmatrix} \sigma \\ \sigma \end{pmatrix} dB_t^1 + \begin{pmatrix} \sigma \\ \sigma \end{pmatrix} dB_t^2 \tag{4.32}$$

Figure 4.5 imposes the conditions $\alpha = a = 1$, $\beta = b = 3$, setting $C_1 = -1$, $C_2 = 3$ in Eq. (4.16).

Figure 4.5a, with volatility but without crosstalk, finds that increasing σ constrains toleration of overload and triggers bifurcation. The simple crosstalk example of Fig. 4.5b appears roughly analogous.

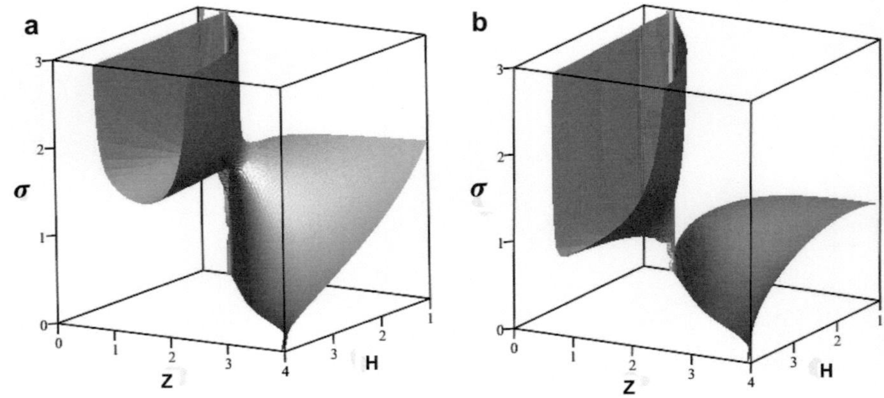

Fig. 4.5 Analog to Fig. 4.2c and d for a mixed model based on Eqs. (4.17), showing the nonequilibrium steady state condition $\langle dL_t \rangle = 0$ from Eq. (4.15). **a** This is first taken under joint, but not cross, stochastic variation in both H and Z, based on Eq. (4.31). Here, $\omega = \sigma$, implying that Clausewitzian fog-and-friction equally burdens delivery of essential resources and multiple signals. Rising σ both constrains toleration of overload and triggers bifurcation. **b** An elementary crosstalk example based on Eq. (4.32)

References

Atlan H., I. Cohen, 1998, Immune information, self-organization, and meaning, International Immunology, 10:711–717.

Brown, R., 1992, Out of line, Royal Institute Proceedings 64:207–243.

Brown, R., P. Higgins, R. Sivera, 2011, Nonabelian algebraic topology: Filtered spaces, crossed complexes, cubical Homotopy Groupoids, EMS tracts in mathematics, Vol. 15.

Cayron, C., 2006, Groupoid of orientational variants, Acta Crystalographica Section A, A62:21040.

Champagnat, N., R. Ferriere, and S. Meleard, 2006, Unifying evolutionary dynamics: From individual stochastic process to macroscopic models, Theoretical Population Biology, 69:297–321.

Cohen, I., 2000, Tending Adam's Garden: Evolving the cognitive immune self, Academic Press, New York.

Cover, T., J. Thomas, 2006, Elements of Information Theory, Second Edition, Wiley, New York.

de Groot, S., P. Mazur, 1984, Nonequilibrium Thermodynamics, Dover, New York.

Dembo, A., O. Zeitouni, 1998, Large deviations and applications, 2nd ed., Springer, New York.

Effros, M., A. Chou, R. Gray, 1994, Variable-rate source coding theorems for stationary nonergodic sources, IEEE Transactions on Information Theory 40:1920–1925.

Feynman, R. Lectures on Computation, Westview Press, New York (2000).

Golubitsky, M., I. Stewart, 2006, Nonlinear dynamics and networks: the groupoid formalism, Bulletin of the American Mathematical Society, 43:305–364.

Gray, R., 1988, Probability, Random Processes, and Ergodic Properties, Springer, New York.

Jackson, D., Kempf, A., Morales, A., 2017. A robust generalization of the Legendre transform for QFT. J. Phys. A 50, 225201.

Khinchin, A., 1957, Mathematical Foundations of Information Theory, Dover, New York.

Laidler, K., 1987, Chemical Kinetics, 3rd ed, Harper and Row, New York.

Leon-Garcia A, L. Davisson, D. Neuhoff, 1979, New results on coding of stationary nonergodic sources, IEEE Transactions on Information Theory, IT-25:137–144.

Maignan, A., and T. Scott, 2016, Fleshing out the generalized Lambert W Function, ACM Communications in Computer Algebra 50:45–60.

Mezo I., G. Keady, 2015, Some physical applications of generalized Lambert functions, arXiv:1505.01555v2 [math.CA] 22 Jun 2015.
Newman, M., 2010, Networks: An Introduction, Oxford University Press, New York.
Pettini, M., 2007, Geometry and topology in hamiltonian dynamics and statistical mechanics, Springer, New York.
Protter, P., 2006, Stochastic Integration and Differential Equations: A new approach, Second edition, Springer, New York.
Schreiber, U., Z. Skoda, 2010, Categorified symmetries, arXiv:1004.2472v1.
Scott, T., R. Mann, R. E. Martinez, 2006, General relativity and quantum mechanics: towards a generalization of the Lambert W-function, Appl. Algebra Engrg. Comm. Comput. 17:41–47.
Shields, P., L. Neuhoff, D. Davisson, F. Ledrappier, 1978, The distortion-rate function for nonergodic sources, The Annals of Probability 6:138–143.
Stewart, I., 2017, Spontaneous symmetry-breaking in a newtwok model for quadruped locomotion, International Journal of Bifurcation and Chaos, 14:1730049 (online).
Tateishi, A., R. Hanel, S. Thurner, 2013, The tranformation groupoid structure of the q-Gaussian family, Physics Letters A 377:1804–1809.
Tegmark, M., 2016, Improved measures of integrated information, PLOS Computational Biology, https://doi.org/10.1371/journal.pcbi.1005123.
Wallace, R., 2005, Consciousness: A Mathematical Treatment of the Global Neuronal Workspace Model, Springer, New York.
Wallace, R., 2011, On the evolution of homochriality, Comptes Rendus Biologies 334:263–268.
Wallace, R., 2012, Consciousness, crosstalk, and the mereological fallacy: an evolutionary perspective, Physics of Life Reviews, 9:426–453.
Wallace, R., 2017, Computational Psychiatry: A systems biology approach to the epigenetics of mental disorders, Springer, New York.
Wallace, R., 2018, New statistical models of nonergodic cognitive systems and their pathologies, Journal of Theoretical Biology 436:72–78.
Wallace, R., 2020a, Signal transduction in cognitive systems: Origin and dynamics of the inverted-U/U dose-response relation, Journal of Theoretical Biology 504: https://doi.org/10.1016/j.jtbi.2020.110377.
Wallace, R., 2020b, Cognitive Dynamics on Clausewitz Landscapes: The control and directed evolution of organized conflict, Springer, New York.
Wallace, R., 2021, Toward a formal theory of embodied cognition, Biosystems https://doi.org/10.1016/j.biosystems.2021.104356.
Wallace, R., 2021b, On Maskirovka: the dynamics of delay in threat recognition, Journal of Defense Modeling and Simulation, https://doi.org/10.1177/15485129211061688.
Wallace, R., 2022, Deception and Delay in Organized Conflict: Essays on the mathematical theory of Maskirovka, Springer, New York.
Wallace, R., 2023, Essays on the Extended Evolutionary Synthesis: Formalizations and Expansions, Sprnger, New York.
Weinstein, A., 1996, Groupoids: unifying internal and external symmetry, Notices of the American Mathematical Association, 43:744–752.
Winkelbauer, K., 1970, On the asymptotic rate of non-ergodic information sources, Kybernetika Cislo 2, Rocnik 6/1970: 127–148.

Chapter 5
Wicked Problems

Abstract Interaction between real-world environments and embodied autonomous cognitive agents within-and-of those environments is channeled by the overwhelming 'riverbank' influences of culture and path-dependent historical trajectory, often intertwined with the burdens of fog, friction, and deadly adversarial intent. These and related factors sculpt conflict at all scales and levels of organization. Using the asymptotic limit theorems of information and control theories, this work formally argues that autonomous intelligence—always and inevitably embodied—must suffer significantly increased probability of failure under such foundational ambiguity, the so-called 'wickedly hard' multiple-option/multiple-intent strategic problems not amenable to straightforward, if often exceedingly difficult, tactical and operational engineering solutions. Problems at the strategic level thus enter realms of subtlety parallel to those that confound the understanding and representation of individual human consciousness, albeit on a differing time scale. In particular, we develop probability models of cognitive dynamics on wickedly hard problems that can be converted to statistical tools for their empirical study and limited control.

> ***Problem understanding and problem resolution are concomitant to each other.***
>
> —Rittel and Webber (1973).

> ***...[T]he narratives of death, suffering, maladministration, malnutrition and disease that have dominated the popular history and public image of the Crimean War have kept us from seeing the war as something else just as significant: a theatre of exuberant technological enterprise.***
>
> —Bektas (2017).

5.1 Introduction

Organized conflict takes place between embodied, cognitive, institutional entities that act on and within—while being acted on by—landscapes of imprecision, uncertainty, delay, resource constraint, and often deadly adversarial intent. Here, we reconsider these matters, extending the perspectives of Wallace (2021, 2022).

R. Wallace, *Hallucination and Panic in Autonomous Systems*, Studies in Computational Intelligence 1217, https://doi.org/10.1007/978-3-031-95207-4_5

Watts (2004) characterized large-scale strategic, as opposed to constrained tactical and operational, military problems as 'wickedly hard'—encompassing exponentially exploding possibilities, compounded by severely limited information about them and similarly limited means to address them. Tactical problems almost always yield to straightforward, if not simple or easy, 'engineering' solutions: "firstist with the mostist", as asserted by the Confederate general Nathan Bedford Forrest.

Rittel and Webber (1973), famously, describe wickedly hard problems in a classic decade of terms:

1. There is no definitive formulation of a wicked problem.
2. Wicked problems have no stopping rule.
3. Solutions to wicked problems are not true-or-false, but good-or-bad.
4. There is no immediate and no ultimate test of a solution to a wicked problem.
5. Every solution to a wicked problem is a 'one-shot operation'; because there is no opportunity to learn by trial-and-error, every attempt counts significantly.
6. Wicked problems do not have an enumerable (or an exhaustively describable) set of potential solutions, nor is there a well-described set of permissible operations that may be incorporated into the plan.
7. Every wicked problem is essentially unique.
8. Every wicked problem can be considered to be a symptom of another problem.
9. The existence of a discrepancy representing a wicked problem can be explained in numerous ways. The choice of explanation determines the nature of the problem's resolution.
10. The planner has no right to be wrong.

The military perspective is markedly different from this work and its civilian progeny. The seminal analysis of Rittel and Webber has spawned an academic policy science consensus—and ideological cottage industry—viewing 'wickedness' as most essentially driven by the need to actively reconcile competing interests, but without much address of imbalances driven by inherent power relations. Western military theory, by contrast, takes the resolution of such competition by prolonged, active, conflict as a given. It is that perspective we adopt here, although perhaps more brutal 'evolutionary' approaches are available (e.g., Wallace 2020b, Chap. 5).

The split in the Western representation of wicked problems has received some comment. Head (2019), for example, writes

> Several critical perspectives have challenged the salience of positivist certainties in social policy and questioned the alluring vision of 'evidence-based' policymaking. Historically, some of these criticisms focused on the power-shaping realities of political and organizational systems. For example, the political-economy critique is that, despite extensive evidence of poverty, public decision-making in practice reflects the interests of power elites, regardless of formal democratic processes such as legislative elections. In this view, the issues chosen for policy attention and the way those issues are defined will generally reflect the structural power of business, their lobbyists and their political representatives... Other critics of positivist social science emphasize the cognitive and sociocultural differences that hinder agreement in a society characterized by divergent values. This cognitive and sociocultural argument is that policy-relevant knowledge about social issues is always plural, not unitary... and therefore, consideration of social issues is ultimately about how divergent perspectives are

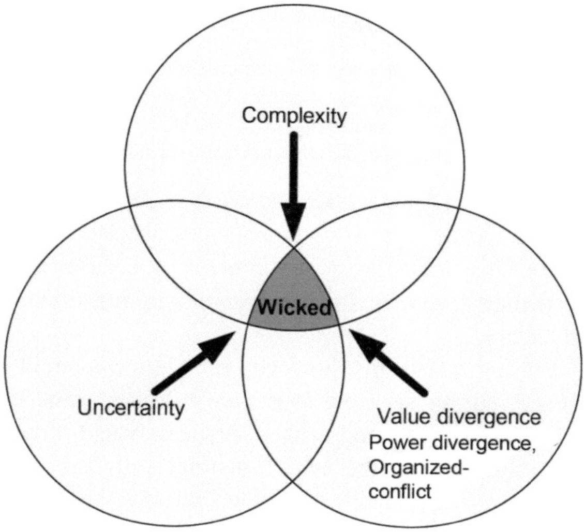

Fig. 5.1 Adapted from Head (2022), Fig. 2.1. Complexity, uncertainty, and 'social modes of expression' in a wickedly-hard problem

> expressed, mobilized and sometimes reconciled... The [mainstream] 'wicked problems' literature is mainly located in this second camp, focusing on how actors in various situations articulate their diverse perspectives about public issues.

Figure 5.1 adapts Fig. 2.1 of Head (2022), explicitly representing, in the lower right lobe, however, the dispersion of values, power relations, and expression of power in organized conflict.

There are different aspects of strategic wickedness. Strachan (2019) identifies a distinction between strategy as the use of battle for the purposes of war and strategy as the use of war for the purposes of policy. This perspective

> ...[F]uels Napoleon's critics, who denounce him for failing to see that the purpose of war was not to create the conditions for the next battle but the capacity to convert war into a lasting peace.

Further, Strachan (2019) warns that

> ...[S]trategy needs to be modest about itself and about what it can deliver. It is, after all, more of an art than a science, and it behooves those who think about it and those who practice it not to be too brazen about its status. Its principles may be guidelines, but, as strategic theorists who are worth their salt have stressed, they are not rules. Precisely because strategy is a pragmatic business it lacks the clarity and purity which strategic theory so often seeks. Strategy has to be reactive as well as predictive, and it must be open to new evidence, whether presented as intelligence or acquired as experience....

Such lack of clarity and purity, as a component of 'wickedness', will be central to our formal analysis.

Gray (2015) describes another fundamental dichotomy that we must attempt to address, that between military tactics and a strategy made up of 'statements' constructed from the 'engineering alphabet' of those tactics:

> The great challenge… is that inherent in the differences in nature between tactics and strategy. These two inclusive concepts are fundamentally distinctive and radically different in meaning… …[A]ny and all [inherently tactical]… military action requires conversion into the different and higher currency of strategy… …[T]actics and strategy… [are]… thoroughly mutually interdependent… …[T]actics… should refer only to military action… while strategy … [regards] the consequences of [military action]….

We will, ultimately, view strategy as representing the output of a 'strategic information source' whose statements, constructed from the tactical alphabet, must certainly have meaning in the context of Clausewitzian fog and friction, but most particularly, also in the context of ongoing dialog with a skilled and dedicated adversary.

Formal treatments, however, remain a minefield, as the experience with 'game theory' during the Cold War shows. Indeed, high level 'true believer' researchers and others are reported to have persistently and actively lobbyed for a thermonuclear first strike against the USSR, justified only by an elementary two-person conflict matrix. A psychopathic Merkwürdig Liebe that might have been taken straight from the classic movie 'Dr. Strangelove'.

With this near-catastrophe in mind, what we seek here is not some simplistic 'elegant' or 'comprehensive' analysis or treatment of wickedly-hard problems, but rather the more modest goal of using the asymptotic limit theorems of information and control theories that constrain cognitive phenomena at any scale or level of organization to build a new class of statistical tools for the study of real-time, real-world data associated with such problems. 'Solving' them involves multiple rounds of feedback and negotiation—often quite brutal—that are subject to sudden 'phase transitions' and other dynamics. These latter, we will argue, can be addressed in formal manners that may provide useful insight in the search for resolution.

We will, in Sect. 5.7, provide a worked-out example of institutional fragmentation under stress, the dynamics of a 'speciation' in the face of selection pressure that, by incubating multiple pathways of response, ultimately confounds selection.

We will, perhaps somewhat surprisingly, argue that the empirical Yerkes-Dodson law (Diamond et al. 2007; Yerkes and Dodson 1908), relating complex task performance to arousal for individual animals under experimental conditions, provides a useful first entry point into formal models of institutional conflict. The YD law has itself, however, been the subject of controversy. As Diamond et al. (2007) put it,

> [The] five-decade-long misrepresentation of Yerkes and Dodson's findings has occurred despite the unambiguous statement by these authors that 'an easily acquired habit may be readily formed under strong stimulation, whereas a difficult habit may be acquired only under relatively weak stimulation.'

Figure 5.2, adapted from Diamond et al. (2007), follows the original Yerkes-Dodson results. The complete idea, rather than simply of an 'inverted-U' dose-response pattern, is that the form of the performance/arousal relation will depend critically on the difficulty of the task. And upon this hangs our tale.

While experimental confirmations of the 'classic' YD law seem relatively rare, there is, as Northoff and Tumati (2019) describe in detail, a vast literature on 'inverted-U' responses under biological signal transduction, where, for example, low doses of

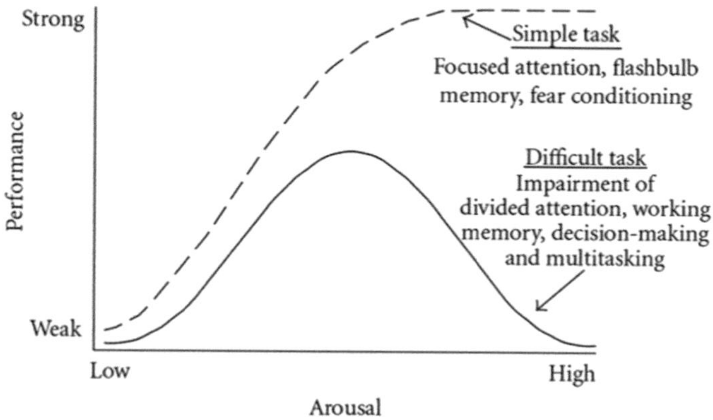

Fig. 5.2 Adapted from Diamond et al. (2007). Two theoretical forms of the Yerkes-Dodson law, for a simple and a difficult task: S-shaped and inverted-U

biomemetic toxins first activate biological mechanisms, and then, at higher dosage, cease to have meaning, and thus trigger decreasing response. A few case histories, across disciplines, include McClintock and Luchinsky (1999), Wallace et al. (2003), Wilken et al. (2000a), Wilson et al. (2000b).

Maturana and Varela (1980) argue that living systems are necessarily cognitive at every scale and level of organization. The inference, then, is that something much like the Yerkes-Dodson relation(s) will be ubiquitous, across cellular processes, individual animal cognition, social function, institutional dynamics, machine cognition, and their many and varied composites, including armies 'enabled' by artificial intelligence.

For systems acting in the real world, all cognition is embodied, purposed, and interacts with an embedding environment that ranges from other cellular components to vast ecological and socioeconomic, and institutional constructs. That is, institutions and real-time machine control systems, are, most particularly, embodied.

The Atlan and Cohen (1998) model of cognition asserts that cognitive process, at its most basic, involves receipt of sensory information from the embedding environment, comparison of that information with an internal, learned or inherited picture of the world, and upon that comparison, choice of an action from a larger repertoire of those available. Such choice, in a formal sense, decreases uncertainty, and reduction of uncertainty implies existence of an information source (Cover and Thomas 2006) 'dual' to that cognitive process. The argument is direct (Wallace 2005, 2022, 2017, 2020a, b).

Any 'internal picture of the world' has a foundational problem. As Manheim (2019) puts it,

> Any intelligence, whether machine learning-based, human, or AI, requires... implicit simplification, since the branching complexity of even a relatively simple game such as Go dwarfs the number of atoms in the universe. Because even moderately complex systems cannot be fully represented... optimization failures... are inevitable. The contrapositive to Conant and

> Ashby's (1970) theorem… is that if a system is more complex than the model, any attempt to control the system will be imperfect.

Imperfect control, then, is perhaps the most fundamental norm for any and all embodied cognitive processes: the real world, with which an embodied cognitive entity interacts, does not often conform to its wishes.

There is a formal characterization of this conundrum, the Data Rate Theorem.

5.2 The Data Rate Theorem

Embodied cognitive systems operating in real time on a real-world environment are inherently unstable under Manheim's contrapositive to the Conant/Ashby Theorem, similar to vehicles driving at night on a twisting, pot-holed roadway. Such vehicles require, in addition to competent drivers, very good headlights, a stable motor, and reliable, responsive steering.

The Data Rate Theorem (DRT), an extension of the Bode Integral Theorem, establishes the minimum rate at which externally-supplied control information must be provided for an inherently unstable control system to remain stable.

The standard argument (e.g., Nair et al. 2007) makes a linear expansion near a nonequilibrium steady state (nss). An n-dimensional vector of system parameters at time t, x_t, determines the state at time $t+1$ according to a relation

$$x_{t+1} = \mathbf{A}x_t + \mathbf{B}u_t + W_t \tag{5.1}$$

$\mathbf{A}$ and $\mathbf{B}$ are fixed $n \times n$ matrices. u_t is the vector of control information, and W_t is an n-dimensional vector of Brownian 'white noise'.

Figure 5.3 projects down onto an irreducible minimum the centrality of any command-and-control structure in the presence of 'noise'—unanticipated difficulties and occurrences that are most often modeled (as we will do here) as an undifferentiated Brownian white noise spectrum, but can become 'colored'—having a shaped power spectrum—by internal structure in more complicated models.

The DRT then states that the minimum rate at which control information H must be provided for system stability is

$$H > \log[|\det[\mathbf{A}^m]|] \equiv H_0 \tag{5.2}$$

det is the determinant of the matrix $\mathbf{A}^m$.

Assuming $m \leq n$, $\mathbf{A}^m$ is the subcomponent of $\mathbf{A}$ having eigenvalues ≥ 1. The right hand side of Eq. (5.2) is characterized in the literature as the rate at which the system generates it's own 'topological information'. See Nair et al. (2007) for details of the standard derivation. Stability is lost if the inequality of Eq. (5.2) is violated. For driving at night, if the headlights fail, or steering becomes unreliable, a twisting roadway cannot be navigated.

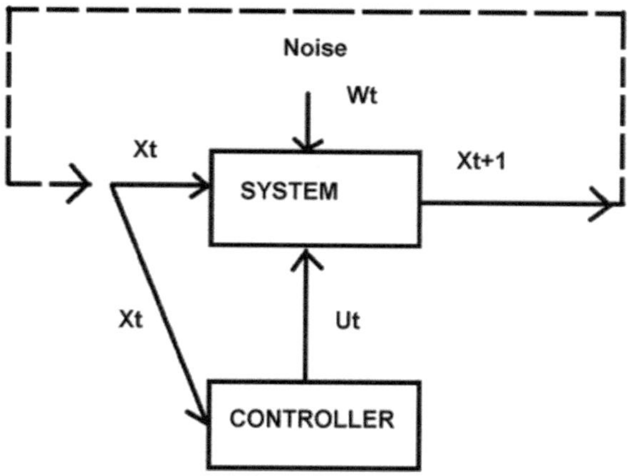

Fig. 5.3 A control system under the constraints of the Data Rate Theorem. The state of the system X, is compared with what is wanted, and a corrective control signal U is sent at an appropriate rate, under conditions of noise represented as W. The rate of transmission of the control signal must exceed the rate at which the inherently unstable system generates its own 'topological information'

French et al. (1999) constructed a neural network model of the Yerkes-Dodson mechanism, concluding

> [Our] results suggest an interpretation of the Yerkes-Dodson law in terms of data compression. The capacity of a network to learn difficult tasks breaks down at high levels of arousal because the computation of network weights associated with the arousal and other inputs strains computational resources. Easy tasks require less computation and hence the network can perform well at higher levels of arousal.

Data compression invokes central matters of information theory, in addition to those of control theory described above. And again, there is a formal context, the Rate Distortion Theorem (Cover and Thomas 2006).

We will develop a basic model that incorporates both kinds of constraint, but we first impose some necessary simplification.

5.3 Scalarizing Driving Parameters

We study system dynamics driven by restrictions on fundamental resources such as information. Indeed, the characterization of internal and external information, and their relations with material resources, in a large sense, requires attention before exploring system dynamics under stress.

We suppose a multicomponent, embodied, cognitive agent enmeshed in a real-world landscape of imprecision in both action and result.

There are, on such landscapes, three essential resources for that agent. The first is the rate at which information can be transmitted between elements of the agent itself, measured by an information channel capacity $\mathcal{C}$. The second is the rate at which 'sensory' information is available to the agent, say $\mathcal{H}$, and the third is the rate at which 'material resources' are available, say $\mathcal{M}$. For neural cells, this is the rate at which metabolic free energy can be supplied. For armed combat, $\mathcal{M}$ is the rate of supply of personnel, ammunition, equipment, and fuel. For driverless cars on intelligent roads, $\mathscr{M}$ is the inverse linear vehicle density, in appropriate combination with a measure of road quality, vehicle responsiveness, and so on.

These three rates will interact, creating some 3×3 correlation matrix analog $\mathbf{Z}$. An $n \times n$ matrix has n scalar matrix invariants that are robust to similarity transformations, constructed according to the standard polynomial relation

$$p(\gamma) = \det[\mathbf{Z} - \gamma \mathbf{I}] = \gamma^n - r_1\gamma^{n-1} + r_2\gamma^{n-2} - \cdots + (-1)^n r_n \tag{5.3}$$

$\mathbf{I}$ is the $n \times n$ identity matrix, det the determinant, and γ a real-valued parameter. The first invariant is the matrix trace, and the last is $\pm$ the matrix determinant.

The n matrix invariants permit construction of a scalar index $Z = Z(r_1, \ldots, r_n)$. The simplest might be $Z = \mathscr{C} \times \mathscr{H} \times \mathscr{M}$. Such scalarization must be appropriate, and there may well be important cross-interactions between the three rates.

This is more subtle than it appears: scalarization permits analysis of a one dimensional system. Expansion of Z into vector or full matrix form leads to multidimensional dynamic equations that are considerably less tractable, and require sophisticated Lie group methods for address (e.g. Wallace 2021b).

5.4 Fog, Friction, and Cognitive Cultural Artifacts

Across the levels and scales of conflict, it is often possible to measure the gap between what has been ordered and what has been performed in real time. Gains, losses, delays, and so on, can be measured and ground up as 'big data' to arrive at an estimate of the gap between what the hand reaches for and what the eye measures as the gap. Such measurement provides formal entry into military tactical and operational dynamics as an engineering discipline, viewing tactics as the alphabet of a strategic message sent along a de-facto operational information channel.

Here, we argue that information theory's Rate Distortion Theorem (RDT) provides a deep perspective on cognition and its failure that directly complements the results of the Data Rate Theorem. The RDT focus is on the difference between what is initially mandated and what is actually observed, viewing cognition-and-control to involve an operational 'channel' along which a 'message' is transmitted. The difference is defined in terms of a scalar distortion measure D. For any such channel, there is a Rate Distortion Function $R(D)$, always convex in D (Cover and Thomas 2006), that determines the minimum channel capacity—a free energy measure from Feynman's (2000) perspective representing an information transmission rate—needed to ensure that the average distortion between intent and execution is less than or equal to D.

The Basic Model

Taking $R(D)$ as a free energy measure, and following the standard view from chemical reaction theory (Laidler 1987), the rate of cognition control above the punctuated threshold arising from the DRT, which we have called H_0, can be characterized—here, for simplicity—as a Boltzmann pseudoprobability:

$$L \equiv \Pr[R > H_0] = \frac{\int_{H_0}^{\infty} \exp[-R/g(Z)]dR}{\int_0^{\infty} \exp[-R/g(Z)]dR} = \exp[-H_0/g(Z)] \qquad (5.4)$$

H_0 is the system limit from the Data Rate Theorem, and $g(Z)$ is a scalar temperature analog depending on the scalar resource rate index Z that must be calculated from first principles. This is not a physical system in which temperature is an independent parameter, and the form of $g(Z)$ must be determined from first principles for each system, depending in detail on the fundamental underlying probability distribution, taken here as the Boltzmann. The argument will be by abduction from Onsager's approximate treatment of nonequilibrium thermodynamics (e.g., de Groot and Mazur 1984).

Probability distributions other than the Boltzmann can be imposed as empirically necessary, particularly as characterized by the observed 'hazard rate' of the system under study,

$$Q(x) = \frac{\rho(x)}{1 - \int_0^x \rho(u)du}$$

$$\rho(x) = \lim_{\epsilon \to 0} Q(\epsilon)\, e^{\int_\epsilon^x \left(-Q(Z) + \frac{\frac{d}{dZ}Q(Z)}{Q(Z)}\right)dZ}$$

For an exponential distribution, with $\rho(m, x) = m\exp[-mx]$, $Q(x) = m$.

Passage through two or more sequential decision states, each described by the same Exponential Distribution, produces an Erlang Distribution of the appropriate order. System cognitive dynamics then diverge markedly from the Boltzmann/Exponential case.

For the Erlang Distribution of order 2, $Q(x) = x/(1 + x)$ tops out with rising x. By contrast, defining $Q \equiv m/(1 + mx)$, which declines monotonically, leads to the Lomax distribution $\rho(m, x) = m/(mx + 1)^2$ that, while 'fat tailed' and without finite first or second moments, parallels the cognitive dynamics of the Exponential/Boltzmann case.

Worked-out examples requiring a numerical solution are given in the Mathematical Appendix for the Gompertz distribution, having different exponentially-increasing hazard rates.

For the sake of simplicity, we proceed using the Boltzmann Distribution.

Treating the denominator of Eq. (5.4) as a statistical mechanical partition function produces an expression for a free energy Morse Function F (Pettini 2007) as

$$\exp[-F(Z)/g(Z)] = \int_0^\infty \exp[-R/g(Z)]\mathrm{d}R = g(Z)$$
$$F(Z) = -\log[g(Z)]g(Z)$$
$$g(Z) = \frac{-F(Z)}{W(n, -F(Z))} \tag{5.5}$$

where $W(n, x)$ is the Lambert W-function of order n, solving the relation solving the relation $W(n, X)\exp[W(n, X)] = X$. Here, for $n = 0, -1$, $W(n, x)$ is real-valued, respectively, over the ranges $x > -\exp[-1]$, $0 > x > -\exp[-1]$. These real-value conditions, it should be noted, explicitly generate the central phase transitions in cognitive dynamics.

Again, Wallace (2025a, b) explore matters driven by a variety of underlying basic probability distributions beyond the Boltzmann.

We next define an associated entropy-analog as the Legendre transform

$$S = -F + Z\mathrm{d}F/\mathrm{d}Z \tag{5.6}$$

and characterize system dynamics in terms of the Onsager approximation from nonequilibrium thermodynamics (de Groot and Mazur 1984). In first order,

$$\mathrm{d}Z/\mathrm{d}t \approx \mu \mathrm{d}S/\mathrm{d}Z = \mu Z\mathrm{d}^2F/\mathrm{d}Z^2 = f(Z) \tag{5.7}$$

where we will set the 'diffusion coefficient' μ equal to 1. Higher order models are 'not difficult' to construct, as empirically necessary (Wallace 2025a).

$f(Z)$ can be described as the *adaptation function* of the cognitive system. Here, we will treat it as an exponential, i.e., $\mathrm{d}Z/\mathrm{d}t = \beta - \alpha Z$, so that $Z(t) \rightarrow \beta/\alpha$, approached at the rate α. Other forms are, of course, possible, although we expect in all cases $Z(t) \rightarrow$ Const..

Taking $F(g(Z))$ as a general expression, and expanding in terms of Eqs. (5.6) and (5.7),

$$F(Z) = \int\int \frac{f(Z)}{Z}\mathrm{d}Z\ \mathrm{d}Z + C_1 Z + C_2 \tag{5.8}$$

Setting $f(Z) = \beta - \alpha Z$, an 'exponential' model,

$$F(Z) = \beta\,(Z\ln(Z) - Z) - \frac{\alpha\, Z^2}{2} + C_1 Z + C_2 \tag{5.9}$$

On 'Clausewitz landscapes' of institutional conflict—'battlefields'—friction, here characterized by $f(Z)$, is a central force, and sets the stage for the temperature analog $g(Z)$, subject to the boundary conditions C_1 and C_2 from Eq. (5.8). $g(Z)$, as an effective 'temperature', will, in turn, drive phase transitions of various forms.

Figure 5.4 examines the cognition rate $L = \exp[-1/g(Z)]$, taking $\partial Z/\partial t = f(Z) = \beta - \alpha Z$. The independent variate, representing 'arousal', is taken to be β

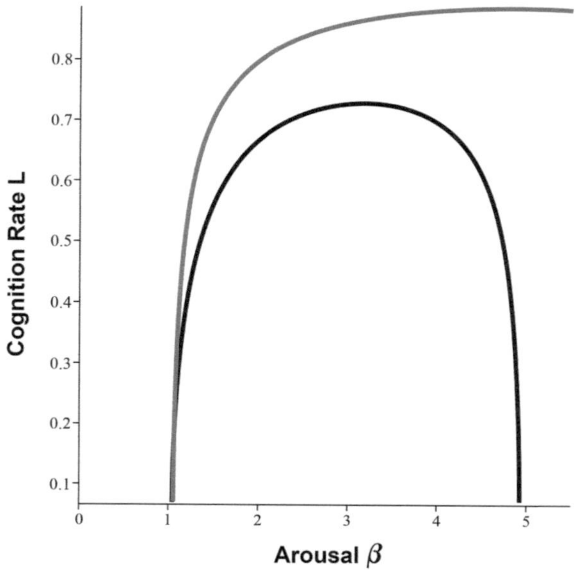

Fig. 5.4 Signal transduction patterns for cognition rates $L(Z) = \exp(-1/g(Z))$, taking $f(Z) = \beta - \alpha Z$, $\alpha = 1.0$. $Z \to \beta/\alpha \equiv Z_{nss}$, letting β vary as the 'arousal' index. $C_1 = -1, \ -5.7$ with $C_2 = 3, \ 8$, for, respectively, the darker and lighter plots. Only for β in a limited range are the two $g(Z)$ functions real-valued, giving the inverted-U for the difficult task. The lighter curve represents a 'simple' task over the same range of arousal β. However, sufficient arousal will crash any cognitive process. Different forms of $f(Z)$, for example $(Z/\alpha)(\log[Z/\beta])^2$, produce similar results

itself. The dark and light plots have the same rate constant $\alpha = 1.0$, with, respectively, $C_1 = -1, \ -5.7$ and $C_2 = 3, \ 8$. We attempt to make the inverted-U cognition rate replicate the pattern of Fig. 5.2b. That is, a 'difficult' problem displays the inverted-U dynamic, while an 'easy' problem simply tops out.

Parameter variations across the model produce a great variety of configurations and relations. Figure 2d of Trofimova et al. (2017) examines something similar.

Other choices for $f(Z)$ produce analogous results.

Heuristically, Clausewitzian fog and friction—'delay in adaptation'—make any cognitive process subject to inverted-U behaviors. However, biological systems having undergone a billion or more years of broad Darwinian and other selection pressures may be more subtle than armies or corporations facing many fewer years of, primarily, far more limited Lamarckian selection.

Fisher Zeros: Temperature as a Complex Number

The third line of Eq. (5.5), via the properties of the Lambert W-function, raises the question of complex numbers for the temperature-analog $g(Z)$ which is, according

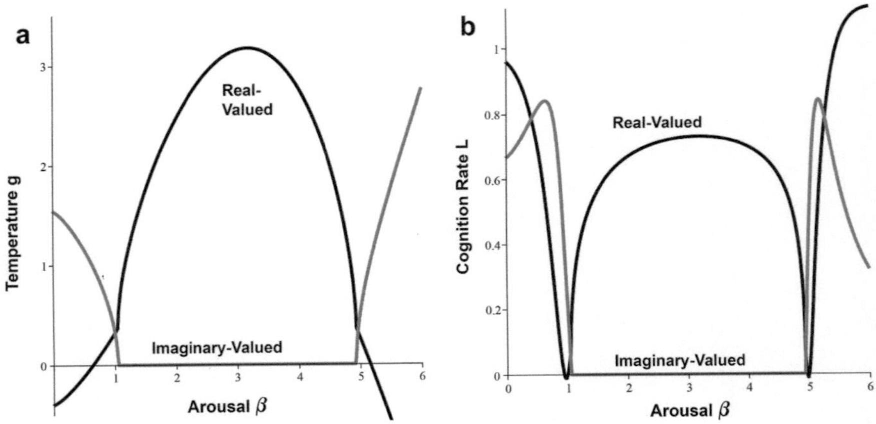

Fig. 5.5 Here we reexamine the sharp inverted-U system of Fig. 5.4, again taking $\alpha = 1$, $C_1 = -1$, $C_2 = 3$. β varies as the index of 'arousal'. **a** both the real and imaginary the values of the temperature analog $g(Z)$, $Z = \beta/\alpha$. The dark trace is the real-valued mode, the lighter the purely imaginary. **b** The corresponding cognition rates. Again, the lighter curve represents the purely imaginary realm. Zeros of the imaginary part of g do indeed indicate phase transition analogs

to the first line of Eq. (5.5), actually the partition function. Physical theory links complex number temperatures in the statistical mechanics of quantum systems with both their phase transitions and their time evolution (e.g., Wei et al. 2014). Ruelle (1964, Sect. 5) provides another perspective. For a full treatment, see Fisher (1965).

We are, in fact, interested in the 'Fisher zeros' of the partition function, focusing on the properties of the Lambert W-function of orders 0 and -1. More specifically, we reexamine the sharp inverted U system of Fig. 5.4, taking $\alpha = 1$, $C_1 = -1$, $C_2 = 3$. Figure 5.5a, however, looks at both the real-values and imaginary-values of the temperature analog/partition function $g(Z)$, $Z = \beta/\alpha$. The dark trace is the real-valued mode, the lighter the imaginary. Figure 5.5b shows how this translates into the cognition rate of Eq. (5.4), again the lighter curve represents the purely imaginary realm. As in the physical model, appearance of 'imaginary temperature' indicates onset of phase transition. Compare with Fig. 6c of Wei et al. (2014). Zeros of the imaginary part of $g(Z)$ do indeed represent phase transition analogs.

Again, the dynamics of cognitive systems are strangely similar to, but essentially different from, those of the physical systems that constitute them.

Adding 'Noise'

Van den Broeck et al. (1994, 1997), Horsthemeke and Lefever (2006), and many others, have explored the role of 'noise' in driving phase transitions in physical systems. Adapting their methods, based on the stability criteria of stochastic differential equations, it is possible to examine 'noise'-driven phase change associated with the embodied interaction of a cognitive agent with a real-world landscape of fog and friction.

The relation $dZ/dt = f(Z) = \beta - \alpha Z$ leads to the stochastic differential equation (Protter 2005)

$$\begin{aligned} dZ_t &= f(Z_t)dt + \sigma Z_t dB_t \\ dZ_t &= (\beta - \alpha Z_t)dt + \sigma Z_t dB_t \end{aligned} \tag{5.10}$$

where the second term imposes 'ordinary' volatility proportional to Z_t, driven at the rate σ. dB_t represents Brownian white noise, the standard starting point for such analyses, which can be extended to 'colored' noise having a nonuniform power spectrum (Protter 2005).

Applying the Ito Chain Rule to Z^2—thereby calculating the nonequilibrium steady state (nss) condition $\langle dZ_t^2 \rangle = 0$—the variance of Z_t can be calculated from the second form of Eq. (5.10) as

$$\langle Z^2 \rangle - \langle Z \rangle^2 = \left(\frac{\beta}{\alpha - \sigma^2/2} \right)^2 - \left(\frac{\beta}{\alpha} \right)^2 \tag{5.11}$$

Variance explodes as $\sigma^2/2 \to \alpha$, independent of β and $Z_\infty = \beta/\alpha$. The rate constant α counts on its own.

Using Eq. (5.10) as the base SDE it is possible to use the Ito Chain Rule on the 'hard' form of the cognition rate $L(Z)$ from Fig. 5.5. A solution can be found for the mean values of L at nonequilibrium steady state, i.e., when $\langle dL_t \rangle = 0$. This convolutes the Lambert W-function with σ, α, and β, and the boundary conditions C_1 and C_2, representing a synergism between the effects of fog, delay, and bifurcation phase transitions. The Ito Chain Rule calculations are quite lengthy but actually routine and easily automated (Cyganowski et al. 2002, Sect. 8.4).

For the 'hard' problem of Fig. 5.5, $\alpha = 1$, $C_1 = -1$, $C_2 = 3$, setting $\beta = 3$—producing the maximum cognition rate—the complicated relation $\langle dL_t \rangle = 0$ can be directly solved for $\sigma(Z)$, as in Fig. 5.6. This gives two solution sets $\{Z, \sigma\}$ and shows critical behavior at $\sigma \approx 1.72$. After that value, the minimum necessary levels of Z far exceed the maximum available. The two equivalence classes of Fig. 5.6 represent a groupoid symmetry-breaking phase transition driven by σ as a 'temperature' analog.

The Lambert W-function(s) in the expression for $\langle dL_t \rangle = 0$ impose necessary conditions driving bifurcation phase transitions on the dynamics—the behavior—of cognitive process under fog and frictional stochasticity:

$$-C_1 Z - Z(-\alpha Z + \beta \ln(Z)) + C_2 - 1/2\, \alpha Z^2 + \beta Z > -\exp(-1) \tag{5.12}$$

Here, bifurcation phase transitions depend critically on noise. 'Noise' and the rate of adaptation—classic Clausewitzian fog and friction—become synergistic with cognitive system phase transitions, generating tortuously complicated and critically

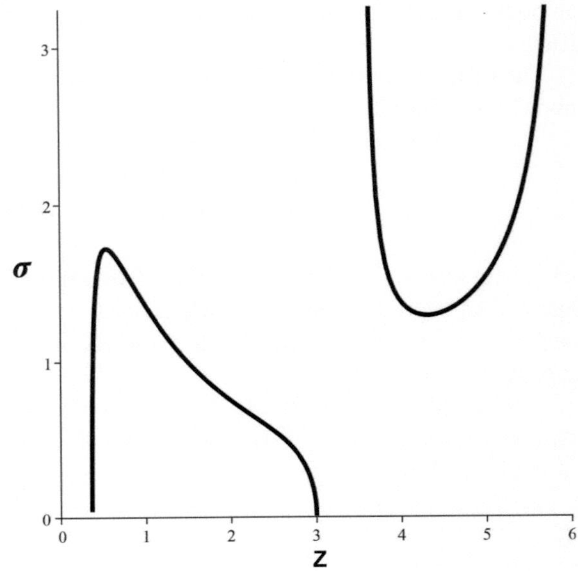

Fig. 5.6 Equivalence classes in $\{Z, \sigma\}$ for the 'hard problem' of Fig. 5.5, defined by the relation $\langle dL_t \rangle = 0$ for cognition rate. This is calculated using the Ito Chain Rule. $\beta = 3$ is fixed at the peak performance point. Cognition fails at $\sigma \approx 1.72$. If σ exceeds that value, the minimum necessary level of Z exceeds the maximum possible value of 3. The two equivalence classes represent a groupoid symmetry-breaking phase transition driven by σ as a 'temperature' analog

unstable patterns of function and failure. These dynamics will become even more complicated when 'noise' has a 'colored' spectral amplitude.

This analysis of Yerkes-Dodson relations in the presence of noise suggests that all cognitive processes at any scale or level of organization, aided by machine cognition or not, can be driven to failure by 'noise', often in highly unexpected and punctuated ways.

Our result has implications for any and all embodied cognitive enterprises burdened with uncertainty, delay, imprecision-in-action, and adversarial intent. Corporations, armies, and real-time AI systems will all likely suffer from such punctuated cognitive failures. In sum, public health systems, emergency rooms, diverless cars, automated stock trading, and AI-partnered military entities, beware.

5.5 The Tyrannies of Time and Resources

Colin S. Gray's master class *Theory of Strategy* (2018) devotes a full tenth of it's text pages to the theme of time, opening with the observation that

> A general theory of strategy can and should advise on the prudent use of time. Even more than with physical geography, as a source of discipline over which they can exercise no control it is a permanent reality for those charged with the strategic well-being of the state... Although time is always a player in strategic affairs... its relevance differs markedly among the distinctive characters of military activity of different kinds and intensity...

Agents of conflict indeed attempt to ride the tiger of time by imposing the functional equivalent of 'anytime algorithms' on their patterns of response. Zilberstein (1996) describes the underlying conceit in these terms:

> ...[Anytime algorithms] give intelligent systems the capability to trade deliberation time for quality of results. This capability is essential for successful operation in domains such as signal interpretation, real-time diagnosis and repair, and mobile robot control. What characterizes these domains is that it is not feasible (computationally) or desirable (economically) to compute the optimal answer...

The essential point is to find a 'good enough' solution in the given time, with the resources actually available during that time.

We examine a hierarchical, chain-like, network constructed of 'tactical', 'operational' and 'command' agents $j = 1 \ldots n$ whose overall quality of response at times T_j, supposing resource rates Z_j, is a scalar measure $Q_i(Z_i, T_i)$, under total constraints $Z = \sum_i Z_i$ and $T = \sum_i T_i$. We assume that $dZ_i/dt = f_i(Z_i)$ as appropriate, typically $f(Z) = \beta - \alpha Z$ or something similar so that $Z(t) \rightarrow$ Const.

We attempt a Lagrangian optimization under environmental 'shadow price' constraints as familiar from economic theory (e.g., Jin et al. 2008; Robinson 1993). Then

$$\begin{aligned}
\mathcal{L} &\equiv \sum_i Q_i(Z_i, T_i) + \lambda\left(Z - \sum_i Z_i\right) + \mu\left(T - \sum_i T_i\right)\\
&\partial\mathcal{L}/\partial Z_i = \partial Q_i/\partial Z_i - \lambda = 0\\
&\partial\mathcal{L}/\partial T_i = \partial Q_i/\partial T_i - \mu = 0\\
&\partial\mathcal{L}/\partial Z = \lambda, \ \partial\mathcal{L}/\partial T = \mu
\end{aligned} \tag{5.13}$$

Since $dZ_i/dT_i = f_i(Z_i)$, a mild abuse of formalism and a little algebra gives

$$\begin{aligned}
\frac{\mu}{\lambda} &= f_i(Z_i(T_i))\\
Z_i(T_i) &= f_i^{-1}\left(\frac{\mu}{\lambda}\right)
\end{aligned} \tag{5.14}$$

Setting $f(Z) = \beta - \alpha Z$,

$$Z_i = \frac{\beta_i - \mu/\lambda}{\alpha_i} \equiv \frac{\beta_i - \omega}{\alpha_i} \tag{5.15}$$

A sufficiently perverse shadow price ratio $\omega \equiv \mu/\lambda$ will drive Z_i in Eq. (5.15) below a threshold, causing system collapse at some critical scale or level of organization.

More detailed models can be found in Wallace (2022, Chap. 1).

Canonical examples are the failure of Wehrmacht supply lines at the Battle of Moscow in December, 1941, the collapse of the Schlieffen Plan at the Battle of the Marne in 1914, and the more recent descent of a pathologically 'lean-and-mean'

Operation Iraqi Freedom into insurgency followed by the rise of ISIS. The contrast to the latter example is, of course, the full-scale Allied occupations of Germany after May, and Japan after August, 1945. Readers will have their own lists.

A different approach to the constraints of time and resources, not depending on $f(Z)$, can be made: we set

$$Q_i(Z_i, T_i) \equiv \frac{Z_i T_i}{\tau_i} \tag{5.16}$$

representing the 'effective rate' of resource delivery to level i under both the shadow price constraints *and the characteristic time constant of the embedding environment of level i* which we write as τ_i.

Elementary calculation finds $Z_i = \mu\tau_i$, $T_i = \lambda\tau_i, T_i = Z_i/\omega$, so that the effective rate of resource supply to level i is constrained as

$$\frac{Z_i T_i}{\tau_i} = \frac{Z_i^2}{\omega\tau_i} < \frac{Z^2}{\omega\tau_i} \to 0 \tag{5.17}$$

where the individual Z_i are fixed and limited by the maximum possible logistic capability $Z = \sum_i Z_i$.

A perverse 'environmental constraint' shadow price ratio ω can, in synergism with a demanding environmental time constant τ_i, according to this model, reduce the effective resource rate below any possible function threshold at some critical level(s) of organization. This is sometimes called the fly-swatter argument.

In addition, the width of the hallucination mode increases as $\approx \sqrt{\tau_i}$.

The inference is that, for 'ecosystem survival', there must be qualitatively different 'species' of autonomous cultural artifacts acting at (and across) each level of organization i so as to—effectively—match the local time constants of the artifact to those of the local embedding environments.

One implication of this condition is that there can never be a single 'Artificial General Intelligence' effective at and across all the interacting scales and levels of organization of systems facing wicked problems.

5.6 Response Time

Pieron's Law

We can do something more with this system, invoking Pieron's Law (Reina et al. 2018) that response time to a signal of strength Z is often found to scale as $T_R \sim Z^{-n}$, $n > 0$. Pieron's Law is broadly consonant with modern neural-specific diffusion decision models (e.g., Ratcliff and McKoon 2008).

We assume that $dZ/dt = f(Z) \geq 0$ is monotonic decreasing and impose 'noise' on Z via the stochastic differential equation

$$dZ_t = f(Z_t)dt + \sigma h(Z_t)dB_t \tag{5.18}$$

where $h(Z)$ is either 1 or else ≥ 0 and monotonic increasing.

Calculating the nonequilibrium steady state for response time as $\langle dZ_t^{-n} \rangle = 0$, using the Ito Chain Rule,

$$\begin{aligned}(n+1)\frac{\sigma^2}{2} &= \frac{f(Z)Z}{h(Z)^2} \equiv Q(Z) \\ Z &= Q^{-1}((n+1)\frac{\sigma^2}{2}) \\ T_R &= \left(\frac{(n+1)\sigma^2/2 + \alpha}{\beta}\right)^n \sim \sigma^{2n}\end{aligned} \tag{5.19}$$

where Q^{-1} represents the inverse function and the latter expression for response time takes $f(Z) = \beta - \alpha Z$ and $h(Z) = Z$.

More generally, T_R wil be monotonically increasing in σ if $Q(Z)$ is sufficiently monotonic decreasing.

On the other hand, some forms of $Q(Z)$ lead to complex values for Z at critical magnitudes of σ. For example, taking $f(Z) = \beta - \alpha Z$ and setting $h(Z) = 1$,

$$T_R = \left(\frac{2\alpha}{\beta + \sqrt{\beta^2 - 2\alpha\,(n+1)\,\sigma^2}}\right)^n \tag{5.20}$$

A sufficient condition that T_R in Eq. (5.20) be real-valued is $\sigma^2 < \beta^2/[2\alpha(n+1)]$. Figure 5.7 shows real and imaginary valued components of $T_R(\sigma)$ taking $n = 2$, $\alpha = 1$, $\beta = 5$. The critical value is $\sigma = 5/\sqrt{6} \approx 2.041$. A nonzero imaginary-valued component characterizes acute instabilities.

The Hick-Hyman Law

The Hick-Hyman Law (Sayood 2028) states that, for some systems, response time is driven by the Shannon Uncertainty of the incoming signal, i.e., the more the uncertainty, the longer it takes to process incoming information.

We model this dynamic for a multicomponent system acting under constraints on both resource rate Z and total available time-to-act T. Each subcomponent receives resources at a rate Z_i for time T_i, and we Lagrange-optimize an information uncertainty measure according to an extended version of Eq. (5.13),

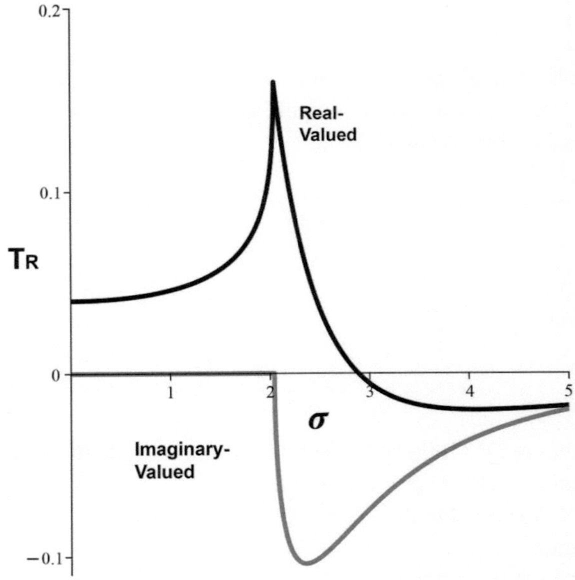

Fig. 5.7 Real and imaginary valued components of $T_R(\sigma)$ from Eq. (5.20), taking $n = 2,\ \alpha = 1,\ \beta = 5$. Appearance of a nonzero imaginary-valued component indexes a Fisher Zero phase transition to instability, here at $\sigma = 5/\sqrt{6} \approx 2.041$

$$
\begin{aligned}
\mathscr{L} &= -\sum_i \frac{Z_i T_i}{ZT} \ln\left(\frac{Z_i T_i}{ZT}\right) + \lambda\left(Z - \sum_i T_i\right) + \mu\left(T - \sum_i T_i\right) \\
\partial\mathscr{L}/\partial Z_i &= -\frac{T_i \ln\left(\frac{Z_i T_i}{TZ}\right)}{TZ} - \frac{T_i}{TZ} - \lambda = 0 \\
\partial\mathscr{L}/\partial T_i &= -\frac{Z_i \ln\left(\frac{Z_i T_i}{TZ}\right)}{TZ} - \frac{Z_i}{TZ} - \mu = 0
\end{aligned}
\tag{5.21}
$$

After some exploration, the relevant expression for response time is found to be

$$
T_R = -\frac{\mu T Z \lambda \ln\left(\dfrac{\mu T Z \lambda}{4\, W\left(-1, -\dfrac{e^{\frac{1}{2}}}{2\sqrt{\frac{1}{\mu T Z \lambda}}}\right)^2}\right)}{4\, W\left(-1, -\dfrac{e^{\frac{1}{2}}}{2\sqrt{\frac{1}{\mu T Z \lambda}}}\right)^2}
\tag{5.22}
$$

where $W(-1, X)$ is the Lambert W-function of order -1, and again we take the 'undetermined multipliers' λ and μ to be shadow prices imposed by environmental and adversarial agencies.

A Fisher-Zero critical phase transition takes place for

$$
\mu T Z \lambda = \frac{4}{\left(e^{\frac{1}{2}}\right)^2 (e)^2} \approx 0.1991
\tag{5.23}
$$

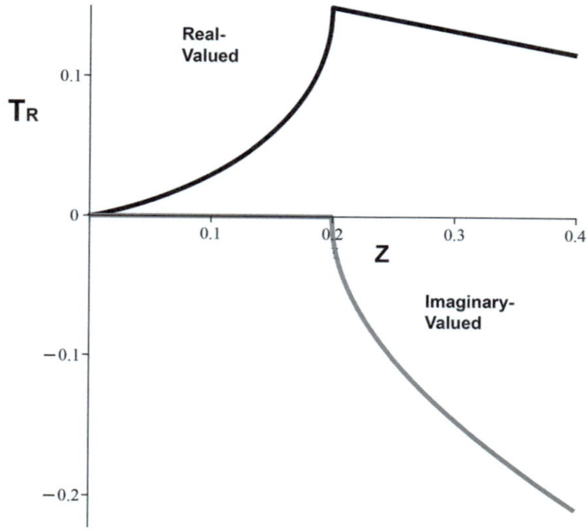

Fig. 5.8 The analog to Fig. 5.7 for the Hick-Hyman Shannon Uncertainty response time in a multicomponent system

Setting $\lambda = \mu = T = 1$, Fig. 5.8 displays real and imaginary-valued components of $T_R(Z)$ and shows qualitatively the same pattern as $T_R(\sigma)$ in Fig. 5.7.

Pieron's Law with Fixed Delay

Taking response time as proportional to Z^{-k}, we explore the destabilizing effect of some fixed delay τ in response, so that $dZ/dt = f(t - \tau)$.

Again taking an exponential model, $dZ/dt = \beta - \alpha Z$, where t is the elapsed time, some clever algebra finds

$$
\begin{aligned}
Z(t) &= \frac{\beta}{\alpha}(1 - \exp[St]) \\
S &= \frac{W(-1, -\alpha\tau)}{\tau} \\
dZ/dt &= S\left(Z - \frac{\beta}{\alpha}\right)
\end{aligned}
\tag{5.24}
$$

where, again, $W(n, X)$ is the Lambert W-function of order n. Exploration for this case finds $n = -1$, so that the function is real-valued only for $-\exp[-1] \le X \le 0$.

Invoking Pieron's Law with $k = 2$, Fig. 5.9 shows real and imaginary components of response time T_R for $\alpha = \beta = 1$, $t = 0.5$. and variable delay τ. Response time first rises slowly with increasing delay, and then more sharply, until a Fisher Zero phase transition at $\alpha\tau = \exp[-1]$.

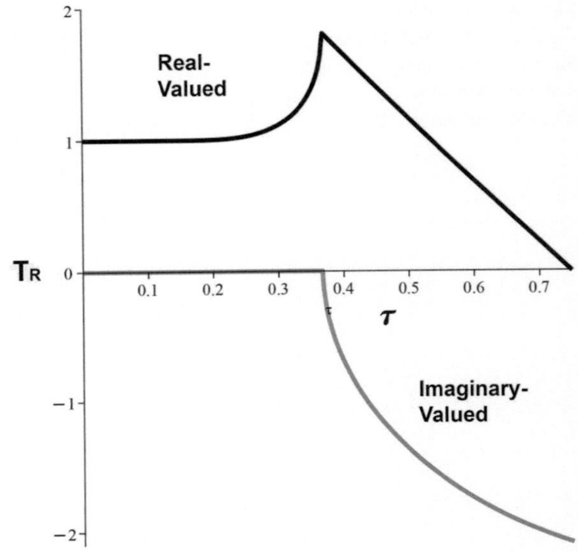

Fig. 5.9 Real and imaginary components of response time T_R for $\alpha = \beta = 1$, $t = 0.5$, and variable delay τ. Here, $k = 2$ in Pieron's Law. Response time first rises slowly with increasing delay, and then more sharply until a Fisher Zero phase transition at $\alpha\tau = \exp[-1]$

Note that the last expression in Eq. (5.24) permits a stochastic treatment according to the first expression of Eq. (5.19).

Pieron's Law with Distributed Delay

If the delay is distributed according to some probability distribution $\rho(\tau)$, then the equation of interest becomes

$$\mathrm{d}Z/\mathrm{d}t = \beta - \alpha \int_0^t \rho(\tau) Z(t-\tau)\mathrm{d}\tau \tag{5.25}$$

Taking $\rho(\tau) = m \exp[-m\tau]$, which has mean value $\langle\tau\rangle = 1/m$, the Laplace Transform method finds

$$Z(t) = \frac{(\alpha m Z_0 + 2\alpha\beta - \beta m)\sinh\left(\frac{t\sqrt{(-4\alpha+m)m}}{2}\right)\mathrm{e}^{-\frac{tm}{2}}}{\sqrt{(-4\alpha+m)\,m}\,\alpha} + \frac{\beta + (Z_0\alpha - \beta)\cosh\left(\frac{t\sqrt{(-4\alpha+m)m}}{2}\right)\mathrm{e}^{-\frac{tm}{2}}}{\alpha} \tag{5.26}$$

where Z_0 is a boundary condition.

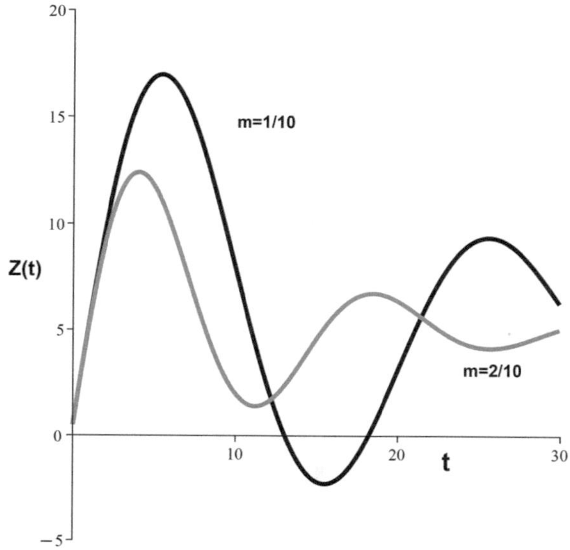

Fig. 5.10 $Z(t)$ for a distributed delay system. Slight changes in the delay distribution can have surprisingly large effect on response time patterns since regimes for which $Z \leq 0$ are inherently unstable, indicating a phase transition to paralysis and thrashing driven by increasing look-back, i.e., rising average τ. The Erlang Distribution, representing sequential passage through the same exponential distribution, gives similar results

Taking $Z_0 = 1/2$, $\alpha = 1$, $\beta = 5$, slightly different values of m produce drastically different patterns of $Z(t)$ as in Fig. 5.10. The region for which $Z \leq 0$ for the $m = 1/10$, $\langle\tau\rangle = 10$ case can be interpreted as a regime of extreme instability—paralysis followed by thrashing—under a Pieron's Law of response time $T_R \propto 1/Z^k$, $k > 0$. Instability is driven by increasing look-back, i.e., rising average τ.

Using the same Laplace Transform method, a similar result emerges for the Erlang Distribution, having the form

$$\rho(k, \lambda, x) = \frac{\lambda^k x^{k-1} \exp[-\lambda x]}{(k-1)!}$$

with $k \geq 1$ an integer and $\lambda > 0$. This distribution characterizes a system filtered in sequence through k independent exponential distributions having the same rate function, a dynamic perhaps typical of decisions within hierarchical institutions and often used to model staged cancer initiation. The pairing is, perhaps, apt.

Enough 'noise' σ in Fig. 5.7, sufficiently large index $\mu T Z \lambda$ in Fig. 5.8, great enough delay τ in Fig. 5.9, or averate distributed delay in Fig. 5.10, impose critical phase transitions determined by the synergisms of external environmental conditions and the intent and abilities of adversaries, often triggering sudden chaotic fragmentation into disconnected subcomponents. Examples might include the post-2003 Iraq 'insurgency', the Northern Mexico Cartel Wars, the collapses of the Whermacht outside of Moscow in 1941 and after Operation Bagration in 1944, or the various precipitate Imperial evacuations from Vietnam and Afghanistan. John Boyd's OODA Loop analysis of tactical and operational enterprise makes much of such dynamics (Osinga 2007).

5.7 'Speciation' as Adaptation to Problem Wickedness

Given fragmentation, we then become interested in fragment size, in a certain sense, seen as a function of the rate at which a system is driven to criticality, i.e., the rate at which a powerful, system-changing 'selection pressure' acts. The approach, abducted from physical theory, is somewhat arcane (Wilson 1971; Zurek 1985, 1996).

We suppose it possible to characterize a 'clump size' R for the system of interest, following the detailed argument of Sect. 6.3 of Wallace (2023, Eq. 6.5). This is not trivial. Then, abducting the Kadanoff approach from physical theory (Wilson 1971), the idea is to invoke a clumping transformation under an external field strength that will be set to zero in the limit. For clumps of size R, given field strength J, and given a measure of coupling strength between individual components ω,

$$
\begin{aligned}
F[\omega(R),\ J(R)] &= Q(R)F[\omega(1),\ J(1)] \\
\chi[\omega(R),\ J(R)] &= \frac{\chi[\omega(1),\ J(1)]}{R}
\end{aligned}
\tag{5.27}
$$

where χ indexes a correlation length across linked information sources.

$Q(R)$ is a 'biological' or 'institutional' renormalization relation that can take various forms, e.g., $R^\delta,\ m\ln(R)+1,\ \exp[m(R-1)/R]$, and so on, so long as $Q(1)=1$ and is monotonic increasing. Physical theory is typically restricted to $Q(R)=R^3$ (Wilson 1971).

Again, see Wallace (2023, Sect. 6.3) for more detailed descriptions of R, ω and related matters.

Assume a critical value of the coupling index ω as ω_C, define $\kappa=(\omega_C-\omega)/\omega_C$, and take $Q(R)=R^\delta$ for $\delta>0$.

Substituting into Eq. (5.27) and taking the limit $J\to 0$, some calculation gives, in first order near ω_C,

$$
\begin{aligned}
F &= \kappa^{\delta/y}F_0 \\
\chi &= \kappa^{-1/y}\chi_0
\end{aligned}
\tag{5.28}
$$

where $y,\ F_0$ and χ_0 are constants arising from the series expansion.

Following Zurek (1985, 1996), let the rate of change of κ remain constant, assigning $|d\kappa/dt|\equiv 1/\tau_\omega$. Argument from physical theory suggests there is a characteristic time constant for the phase transition, $\tau=\tau_0/\kappa$ such that, if changes in κ take place on a time scale longer than τ for any given κ, the correlation length $\chi=\chi_0\kappa^{-s}$, $s=1/y$, will be in equilibrium with internal changes, and result in a very large fragment in R-space.

Zurek next argues that a critical freeze-out time $\hat{t}$ occurs at a system time $\hat{t}=\chi/d\chi/dt$ such that $\hat{t}=\tau$. Evaluating $d\chi/dt$, remembering the definition $d\kappa/dt=1/\tau_\omega$, then

$$\frac{\chi}{|d\chi/dt|} = \frac{\kappa\tau_\omega}{s} = \frac{\tau_0}{\kappa}$$
$$\kappa = \sqrt{s\tau_0/\tau_\omega} \tag{5.29}$$

Substitution of this value for κ into the expression for correlation length χ gives the expected size of fragments in R-space, which we call $d(\hat{t})$, as

$$\mathrm{d} \approx \chi_0 \left(\frac{\tau_\omega}{s\tau_0}\right)^{s/2} \tag{5.30}$$

with $s = 1/y > 0$.

Consequently, the more rapidly ω approaches the critical value ω_C, the smaller is τ_ω and the smaller and more numerous are the R-space fragments.

That is, the rate of 'selection pressure' determines the size of the 'fragments' acting at that rate. The faster, the smaller. Hence the scale sequence Tactical → Operational → Strategic.

An exchange with the Perplexity AI system (Perplexity 2025) can be applied to the consequences for ultimate strategic enterprise of an adversary's fragmentation into multiple independent agents. What emerges is, adopting the term used by the AI, a 'Paradox of Precision':

> AI systems reducing tactical friction often increase strategic uncertainty through:
>
> $$\text{System Stability} \propto \frac{\text{Information Rate}}{\text{Adversarial Innovation Rate}}$$
>
> This creates an arms race where defensive AI updates lag behind offensive adaptation... While AI enhances localized decision-making speed, [recent] work confirms that Clausewitzian dynamics persist through three mechanisms: adversarial rule-changing..., emergent complexity in system interactions..., and the inherent unpredictability of strategic reasoning... These factors ensure that 'disembodied oracles' remain vulnerable to strategic surprise, no matter their technical sophistication...

The more numerous the distinct adversarial fragments, the higher the 'Adversarial Innovation Rate.'

The argument can be interpreted as a slightly different version of that leading to Eq. (5.17), where τ_i, the local 'adversarial' time constant, may, if initiative lies with the adversary, be much longer than time constraints imposed by that adversary on the defending institution.

In the Mathematical Appendix we reprise the argument leading to Eq. (5.17) from the perspective of the Lagrangian of Eq. (5.13), using the Shannon-Hartley Theorem as a starting point.

As the Wehrmacht found on the Eastern Front, The USA in Korea after Inchon, the French and Americans in Vietnam, the USA in the Occupation of Iraq, England, the USSR and the USA in Afghanistan, 'reduction of tactical friction' via advanced technologies has limited ability to mitigate the long-term impacts of profound cultural blindness synergistic with outright strategic incompetence that, in the cases of Russia and the USA, has become deeply institutionalized.

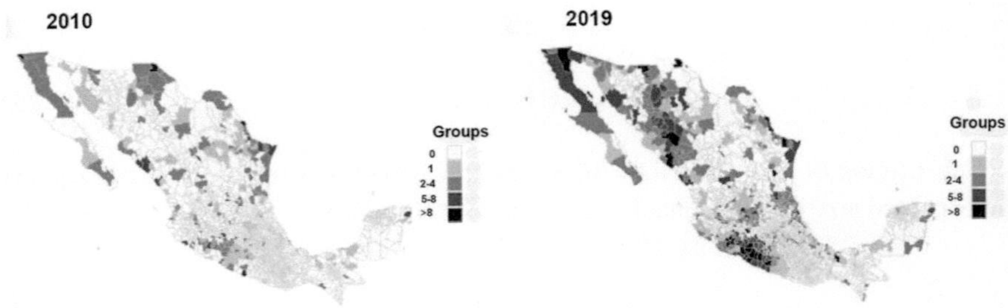

Fig. 5.11 Adapted from Fig. 1 of Esberg (2025). Fragmentation of Mexican criminal groups between 2010 and 2019, constituting an ongoing 'speciation' response to 'drug war' selection pressures

A classic example of adaptation-by-fragmentation can be found in the response of Mexican drug cartels to the 2006 'war on drugs'. As Müller-Ibold (2024) writes,

> The sustained high levels of conflict between organised crime groups follow the fragmentation of Mexican cartels after the launch of the war on drugs in 2006 and the implementation of the kingpin strategy, which sought to combat criminal organisations by targeting their leadership. While drug trafficking operations were formerly controlled by a handful of dominant organisations, in several instances the kingpin strategy contributed to such organisations breaking up into smaller but more violent groups. According to [Uppsala Conflict Data Program] records, the number of criminal organisations involved in at least one death, increased from just four in 2007 to 25 in 2022.

As Esberg (2025) put the matter,

> ...[T]raditional drug war strategies, like leadership removal and interdiction, contributed to the atomization of criminal conflict. This is not unique to Mexico: the decentralization and disintegration of larger criminal groups has happened in states like Colombia and Haiti. Understanding when and why these smaller organizations emerge or expand can thus inform debates about addressing criminal conflict.

Figure 5.11, adapted from Esberg (2025), displays the 'speciation' of criminal organizations in Mexico between 2010 and 2019 under 'drug war' selection pressure.

5.8 Discussion

Cognitive systems acting in, and acted on by, the real world in real time, are inherently embodied, and, we argue, characterized by complex temperature analogs driving dynamics far more subtle than are analogs familiar from physical theory. 'Temperature' itself, apparently, becomes an order parameter. Embodied cognition is, it seems, literally another world, an alien landscape that we experience from our own behaviors, but is much different from the 'simple' physical world of ice, water, and steam, of solid rock and flowing lava, of sequential 'big bang' condensations and the evolution of galaxies. Embodied cognitive enterprise lives in these two worlds, and, to survive, must learn how to navigate both within and between them, as modulated, among humans, by both embedding cultures and the path-dependent burdens of history.

Such navigation, beyond the most elementary see-and-grasp, is often wicked in the sense of Rittel and Webber (1973), both in terms of the 'simple' necessity of balancing multiple competing interests, and in terms of active conflict/conversation with similarly cognitive adversaries. In essence, a contending entity's strategic cognitive information source must compose a sufficiently relevant message to be sent along operational channels using a basic, and usually rigid, tactical alphabet.

There are surprises, and one is the set of generalized Yerkes-Dodson relations that necessarily mark cognitive process under real-world uncertainties, delays, and imprecisions, what Western military science calls Clausewitzian fog and friction (e.g., Watts 2004; Wallace 2020a, b). While difficult tactical/operational problems, where one has available measures of optimization—the hand reaches, the eye measures the gap—seem addressable by engineering methods and variants of control theory, the more and most complex wicked problems confronting institutional strategic decision making appear mired in many of the same conundrums that confront current theories of individual consciousness.

This is perhaps not entirely unexpected to those who have grappled with such matters. After all, organized conflict is an intensely human enterprise, at all scales and levels of organization, expressing the full spectrum of culturally-driven and constrained expressions of human consciousness. While institutions are not conscious, a term restricted to 100 ms time constants in neural systems, many of the 'tunable global workspace' dynamics described in consciousness theory (Wallace 2022b) appear to generalize through the crosstalk linking both institutional work groups and 'voting' adversaries.

Straightforward analysis suggests that some extended version of the Yerkes-Dodson relations will affect and afflict all embodied cognitive phenomena, from cellular signal transduction through institutional conflict and collapse, and the forthcoming, manifold bizarre misfortunes that will be inflicted upon us by the successful marketing of culturally-embedded artificial intelligence for the control of real-time critical systems.

The physical environment and embodied cognition within and of that environment, engage in a dynamic synergism that, for all human enterprise, includes the powerful determinants of culture and path-dependent historical trajectory acting as 'riverbanks' to channel cognitive enterprise.

Push hard enough, and, we claim, something like Figs. 5.4 through 5.10 must emerge for any and all embodied cognitive processes, 'enhanced' by artificial intelligence or not. The formal descriptions of such syncretisms developed here may provide novel insight.

In sum, this modeling exercise suggests—but, constrained as are all such mathematical models of complex social and biological phenomena, does not prove—that embodied cognition will fail under sufficient burdens of fog, friction, and adversarial intent. These remain empirical questions subject to experimental and observational exploration and exploitation. While the inevitability of such failure is, perhaps, painfully 'obvious', it has been reached by formal methods that may be converted to validated statistical models for the analysis of real time data. Such work remains to be done, but might both extend our understanding of wicked problems, and permit added, if inevitably limited, control over them.

Mathematical Appendix

Hazard Rate Tuning

For the Gompertz distribution the hazard rate increases exponentilly, first taken as $Q(x) = \exp[x] + 1$, leading to

$$\rho(x) = \frac{e^{-e^x+1}\left(e^x+1\right)}{e^x}$$

$$\rho(F/g) = \frac{e^{-e^{\frac{F}{g}}+1}\left(e^{\frac{F}{g}}+1\right)}{e^{\frac{F}{g}}} = g$$

$$L = e^{-\frac{e^{\frac{1}{g}}g-g+1}{g}} \tag{5.31}$$

where we have set $R_0 = 1$ in the expression for L. Characteristically, $\rho(x)$ has a sharply-declining reverse-J form.

The second expression must be solved numerically for $g(F)$, taking $F(\beta)$ as in Eq. (5.9), with $\alpha = 1,\ C_1 = -1,\ C_2 = 3$ and β as the index of arousal. Numerical expressions for the two nontrivial solutions to the second expression are shown in Fig. 5.12a, and the corresponding cognition rates in Fig. 5.12b, a surprisingly straightforward inverse-U Yerkes-Dodson result similar to Fig. 5.5b for the Boltzmann Distribution.

By contrast, taking $Q(x) = \exp[x] - 1$ leads, using the same numerical methods, to the narrow 'spiked' system of Fig. 5.13: The inverse-U disappears, and the system is highly pathological.

Taking $Q(x) = \exp[x]$ leads again to a double-spiked cognition rate similar to Fig. 5.13.

Clearly, 'hazard rate tuning' is likely to be important in the cognitive dynamics of the systems studied by these methods.

Shannon-Hartley 'Bandwidth' Fragmentation

Given the limits on Z_i and T_i in the multicomponent system of Eqs. (5.13) and (5.17), we explore maximizing a total channel capacity C according to the Shannon-Hartley

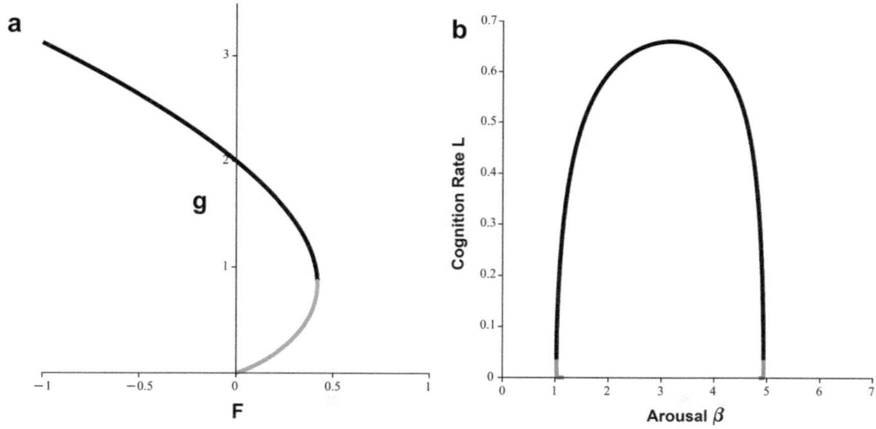

Fig. 5.12 Gompertz Distribution defined by the hazard rate $\exp[x] + 1$, using numerical methods. **a** There are two nontrivial numerical solutions for $g(F)$. F is as in Eq. (5.9) and Fig. 5.5b. **b** Corresponding cognition rates versus arousal, a Y-D Effect similar to Fig. 5.5b. This is a special case

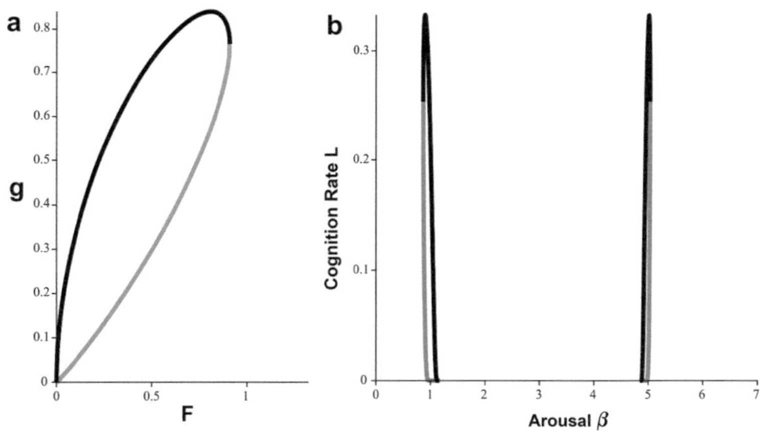

Fig. 5.13 Imposing the hazard rate $\exp[1] - 1$, using numerical methods. **a** $g(F)$ versus F. Again, there are two nontrivial solutions. **b** The cognition rate with F defined as in Fig. 5.12. The inverse-U of the Y-D Effect collapses into two pathologically narrow spikes. 'Hazard rate tuning' is likely to be an important factor in the systems studied by these methods

law that individual subsystem channel capacities will be constructed via Eq. (5.16) as

$$
\begin{aligned}
C_i &= B_i \ln\left(1 + \frac{S_i}{N_i}\right) \\
&= B_i \ln\left(1 + \frac{Z_i T_i/\tau_i}{\sigma_i^2}\right)
\end{aligned}
\tag{5.32}
$$

where, in the first line, classically, S_i is the signal strength, N_i the 'noise' and B_i is the 'available bandwidth' at subsystem i.

In the second expression, τ_i is a characteristic 'environmental' functional time limit for subsystem i, and T_i the available time slice. Z_i is the local composite 'resource rate' scalarization, and σ_i is the local 'noise' burden, leading to the Lagrangian

$$\mathscr{L} = \sum_i C_i + \lambda \left(Z - \sum_i Z_i \right) + \mu \left(T - \sum_i T_i \right) \tag{5.33}$$

where λ and μ are again global environmental shadow price constraints.

Under optimization, $\partial C_i / \partial Z_i = \lambda$ and $\partial C_i / \partial T_i = \mu$, leading to

$$\begin{aligned} \frac{B_i T_i}{\tau_i \sigma_i^2 \left(1 + \frac{T_i Z_i}{\tau_i \sigma_i^2} \right)} &= \lambda \\ \frac{B_i Z_i}{\tau_i \sigma_i^2 \left(1 + \frac{T_i Z_i}{\tau_i \sigma_i^2} \right)} &= \mu \end{aligned} \tag{5.34}$$

These expressions can be solved to give

$$Z_i = \frac{1}{2\lambda} \left(B_i + \sqrt{B_i^2 - 4\lambda\mu\tau_i\sigma_i^2} \right) \tag{5.35}$$

leading to the necessary stability condition on local bandwidth as

$$B_i \geq 2\sqrt{\lambda\mu\tau_i}\sigma_i \tag{5.36}$$

Local bandwidths below such limits will trigger instability in what appears to be a close analog to the Data Rate Theorem.

References

Atlan H., I. Cohen, 1998, Immune information, self-organization, and meaning, International Immunology, 10:711–717.

Bektas, Y., 2017, The Crimean War as a technological enterprise, Notes and Records of the Royal Society, 71:233–262.

Bigsby, R., R. Chapin, G. Daston, B. Davis, J. Gorski, L. Gray, K. Howdeshell, R. Zoeller, F. Vom Saal, 1999, Evaluating the effects of endocrine disruptors on endocrine function during development, Environmental Health Perspectives, 107(Suppl. 4), 613–618.

Brown, R., 1992, Out of line, Royal Institute Proceedings 64:207–243.

Cayron, C., 2006, Groupoid of orientational variants, Acta Crystalographica Section A, A62:21040.

Champagnat, N., R. Ferriere, and S. Meleard, 2006, Unifying evolutionary dynamics: From individual stochastic process to macroscopic models, Theoretical Population Biology, 69:297–321.

Conant R., W. Ross Ashby, 1970, Every good regulator of a system must be a model of that system, International Journal of Systems Science 1:89–97.
Cover, T., J. Thomas, 2006, Elements of Information Theory, Second Edition, Wiley, New York.
Cyganowski, S., P. Kloeden, J. Ombach, 2002, From Elementary Probability to Stochastic Differential Equations with MAPLE, Springer, New York.
de Groot, S., P. Mazur, 1984, Nonequilibrium Thermodynamics, Dover, New York.
Dembo, A., O. Zeitouni, 1998, Large deviations and applications, 2nd ed., Springer, New York.
Diamond, D., A. Campbell, C. Park, J. Halonen, P. Zoladz, 2007, The temporal dynamics model of emotional memory processing.... Neural Plasticity, https://doi.org/10.1155/2007/60803.
Esberg, J., 2025, Criminal fragmentation in Mexico, Political Science Research and Methods, https://doi.org/10.1017/psrm.2025.4.
Feynman, R. 2000, Lectures on Computation, Westview Press, New York.
Fisher, M. E., 1965, Lectures in Theoretical Physics [ed. Brittin W. E.], University of Colorado Press, Boulder.
French, V., E. Anderson, G. Putman, T. Alvager, 1999, The Yerkes-Dodson law simulated with an artificial neural network, Cognitive Systems 5:136–147.
Gray, C.S., 2015, Tactical Operations for Strategic effect: The Challenge of Currency Conversion, JSOU Press, MacDill Air Force Base, FL.
Gray, C.S., 2018, Theory of Strategy, Oxford University Press, New York.
Head, B., 2019, Forty years of wicked problems literature: forging closer links to policy studies, Policy and Society, 38:180–197.
Head, B., 2022, Wicked Problems in Public Policy: Understanding and responding to complex challenges, Palgrave-Macmillan Open Access, New York.
Horsthemeke, W., R. Lefever, Noise-induced Transitions, 2006, Vol. 15, Theory and Applications in Physics, Chemistry, and Biology, Springer, New York.
Jin, H., Z. Hu, and X. Zhou, 2008, A convex stochastic optimization problem arising from portfolio selection, Mathematical Finance 18:171–183.
Khinchin, A.,1957, Mathematical Foundations of Information Theory, Dover, New York.
Laidler, K., 1987, Chemical Kinetics, Third Edition, Harper and Row, New York.
Manheim, D., 2019, Multiparty dynamics and failure modes for machine learning and artificial intelligence, Big Data and Cognitive Computing, 3: https://doi.org/10.3390/bdcc3020021.
Marusich, L., J. Bakdash, E. Onal et al., 2016, Effects of information availability on command-and-control decision making: performance, trust, and situation awareness, Human Factors, 58:301–321.
Maturana, H., F. Varela, 1980, Autopoiesis and Cognition. Dordrecht: Reidel Publishing Company.
McClintock, P., D. Luchinsky, 1999, Glorious noise, The New Scientist, 161(2168), 36–39.
Müller-Ibold, M., 2024, The Organized Crime Landscape in Mexico, Vision of Humanity, https://www.visionofhumanity.org/the-organised-crime-landscape-in-mexico/
Nair, G., F. Fagnani, S. Zampieri, R. Evans, 2007, Feedback control under data rate constraints: an overview, Proceedings of the IEEEE 95:108–137.
Northoff, G., S, Tumati, 2019, 'Average is good, extremes are bad' – Non-linear inverted U-shaped relationship between neural mechanisms and functionality of mental features, Neuroscience and Biobehavioral Reviews, 104:11–25.
Osinga, F., 2007, Science, Strategy and War: The strategic theory of John Boyd, Routledge, London.
Perplexity AI, 2025, Exchange available from the author rodrick.wallace@gmail.com. Accessed 2/6/2025.
Pettini, M., 2007, Geometry and topology in hamiltonian dynamics and statistical mechanics, Springer, New York.
Protter, P., 2005, Stochastic Integration and Differential Equations: A new approach, Second edition, Springer, New York.
Ratcliff, R., G. McKoon, 2008, The diffusion decision model: Theory and data for two-choice decision tasks, Neural Computation 20:877–922.

Reina, A., T. Bose, V. Trianni, J. Marshall, 2018, Psychophysical laws and the superorganism, Scientific Reports, 8:4387, https://doi.org/10.1038/s41598-018-22616-y.
Rittel, H., M. Webber, 1973, Dilemmas in a general theory of planning, Policy Sciences, 4:155–169.
Robinson, S., 1993, Shadow prices for measures of eectiveness II: General model, Operations Research 41:536–548.
Ruelle, D., 1964, Cluster property of the correlation functions of classical gases, Reviews of Modern Physics, April, 580–584.
Sayood, K., 2028, Information theory and cognition: A review, Entropy, 20:706, https://doi.org/10.3390/e20090706.
Stewart, I., 2017, Spontaneous symmetry-breaking in a newtwok model for quadruped locomotion, International Journal of Bifurcation and Chaos, 14:1730049 (online).
Strachan, H., 2019, Strategy in theory; strategy in practice, Journal of Strategic Studies 42:171–190.
Trofimova, I., T. Robbins, W. Sulis, J. Uher, 2017, Taxonomies of psychological individual differences: biological perspectives on millennia-long challenges, Philosophical Transactions B, 373:20170152.
Van den Broeck, C., J. Parrondo, R. Toral, 1994, Noise-induced nonequilibrium phase transition, Physical Review Letters, 73:3395–3398.
Van den Broeck, C., J. Parrondo, J., R. Toral, R. Kawai, 1997, Nonequilibrium phase transitions induced by multiplicative noise, Physical Review E, 55:4084–4094.
Wallace, D., R. Wallace, V. Rauh, 2003, Community stress, demoralization, and body mass index: evidence for social signal transduction, Social Science and Medicine 56:2467–2478.
Wallace, R., 2005, Consciousness: A Mathematical Treatment of the Global Neuronal Workspace Model, Springer, New York.
Wallace, R., 2012, Consciousness, crosstalk, and the mereological fallacy: an evolutionary perspective, Physics of Life Reviews, 9:426–453.
Wallace, R., 2017, Computational Psychiatry: A systems biology approach to the epigenetics of mental disorders, Springer, New York.
Wallace, R., 2020a, On the variety of cognitive temperatures and their symetry-breaking dynamics, Acta Biotheoretica, https://doi.org/10.1007/s10441-019-09375-7.
Wallace, R., 2020b, Cognitive Dynamics on Clausewitz Landscapes: The control and directed evolution of organized conflict, Springer, New York.
Wallace, R., 2021, Embodied cognition and its pathologies: The dynamics of institutional failure on wickedly hard problems, Communications in Nonlinear Science and Numerical Simulation, 95:105616.
Wallace, R., 2021b, How AI founders on adversarial landscapes of fog and friction, Journal of Defense Modeling and Simulation https://doi.org/10.1177/1548512920962227.
Wallace, R., 2022, Essays on Strategy and Public Health: The systematic reconfiguration of power relations, Springer, New York.
Wallace, R., 2022b, Consciousness, Cognition and Crosstalk: the evolutionary exaptation of nonergodic groupoid symmetry-breaking, Springer, New York.
Wallace, R., 2023, Essays on the Extended Evolutionary Synthesis: Formalizations and expansions, Springer, New York.
Wallace, R., 2023b, On 'Machine Consciousness', Journal of Artificial Intelligence and Consciousness, 10:125–148, https://doi.org/10.1142/s2705078522500187.
Wallace, R., 2025b, Mathematical Essays on Embodied Cognition: Insights from information and control theories, Springer, New York.
Wallace, R., 2025a, Sensory or intelligence data compression can drive the Yerkes-Dodson Effect, Symmetry, 17(2), 235; https://doi.org/10.3390/sym17020235.
Watts, B, 2004, Clausewitzian Friction and Future War, Revised Edition, Institute for National Strategic Studies, National Defense University, Washington D.C.
Wei, B., S. Chen, H. Po, R. Liu, 2014, Phase transitions in the complex plane of physical parameters, Scientific Reports, 4:5202, https://doi.org/10.1038/srep05202.

Weinstein, A., 1996, Groupoids: unifying internal and external symmetry, Notices of the American Mathematical Association, 43:744–752.

Wilken, J., B. Smith, K. Tola, M. Mann, 2000, Trait anxiety and prior exposure to non-stressful stimuli: Effects on psychophysiological arousal and anxiety. International Journal of Psychophysiology, 37, 233–242.

Wilson, J., N. Nugent, J. Baltes, S. Tokunaga, T. Canic, B. Yourn, E. Bellinger, D. Delac, G. Golston, D. Henderson, 2000. Effects of low doses of caffeine on aggressive behavior of male rats, Psychological Reports, 86:941–946.

Wilson K., 1971, Renormalization group and critical phenomena. I Renormalization group and the Kadanoff scaling picture. Physics Reviews B 4:317483.

Yerkes, R., J. Dodson, 1908, The relation of strength of stimulus to rapidity of habit-formation, Journal of Comparative Neurology and Psychology, 18:459–482.

Zilberstein, S., 1996, Using anytime algorithms in intelligent systems, AI Magazine 17:73–83.

Zurek,W., 1985, Cosmological experiments in superfluid helium? Nature, 317:505–508.

Zurek, W., 1996, The shards of broken symmetry, Nature, 382:296–298.

Chapter 6
High Frequency Trading and Its Discontents

Abstract Long-persisting cognitive biological sub-systems, higher animal species, and social/institutional cognitive agents and agencies, have always closely paired cognition with regulation. Blood pressure, while necessarily responsive to physiological demand, must be kept within limits. The cognitive immune response must be prevented from excessive self-attack. The stream of consciousness must be held within riverbanks useful both to the individual and the social groupings within which each individual must operate. Institutional function must be held between the Scylla of inherited or learned Doctrine and the Charybdis of Law and Culture. Current HFT agents that pair high-speed trading algorithms with AI fluctuation prediction seem to have attained levels of effective embodiment and technical sophistication that far outstrip even the most advanced realms of air combat. Using formal methods, we find that, absent draconian regulation, increasingly sophisticated HFT-AI systems will, both individually and collectively, increasingly engage in highly punctuated 'sorcerer's apprentice' phase transitions characteristic of both intelligence and consciousness.

Tactics without strategy is the noise before defeat.

—Sun Tzu.

[T]he psychological actions of drivers make traffic different from any other flow.

—Orosz et al. (2010).

R. Wallace, *Hallucination and Panic in Autonomous Systems*, Studies in Computational Intelligence 1217, https://doi.org/10.1007/978-3-031-95207-4_6

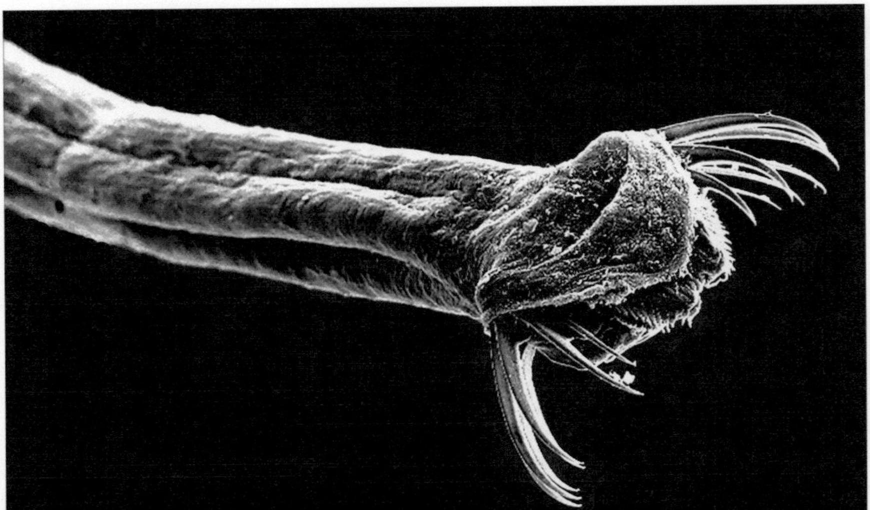

A Sagitta arrow worm with bristles in position for grasping prey.
Image by Zatelmar CC BY-SA 3.0,
https://commons.wikimedia.org/w/index.php?curid=2777766

6.1 Introduction

High frequency stock trading (HFT) algorithms, increasingly linked to artificial intelligence, currently complete some 70% of stock trades in the USA (IMF 2024; Turiel and Aste 2022). These composite entities, like AI-driven air and surface vehicles and emerging weapons systems—and like the arrow worm of the chapter frontispiece—are 'embodied', that is, rapidly affecting and adjusting to an embedding environment. Our central conceit will be that such systems, both individually and collectively via network 'crosstalk' interactions, will—absent draconian regulation—increasingly engage in highly punctuated and apparently idiosyncratic patterns of phase transition, particularly under heavy 'arousal', in a certain sense.

While arrow worms do not engage in collective action, the voracity of these 2–3 cm 'autonomous agents' that have been with us since Cambrian times has made them the second-most common form of zooplankton. Indeed, plankton samples placed in petri dishes must be rapidly euthanized with formaldehyde lest the arrow worms eat everything else, and then cannibalize themselves. In the larger ecosystem, arrow worms are constrained by the predation of higher order creatures. In an inherently unstable AI-enabled stock trading system, what will be the higher order predator needed for stabilization?

To examine this question, we begin by reconsidering recent relatively narrow examples of AI success from larger perspectives.

Demis Hassabic and John Jumper shared part of the 2024 Nobel Prize in Chemistry for using artificial intelligence to solve the 50-year-old problem of predicting protein structures from amino acid sequences (Anfinsen 1973). Their work represents—specifically—solution to the in vitro protein folding problem.

However, as Outerial et al. (2022) argue, predicting the native state of a protein has long been considered only an entry-level gateway for understanding in vivo protein folding dynamics. That is, solution in vitro does not illuminate the physics of how proteins dynamically fold into their equilibrium structure and may just be accurate predictors of the final state. Thus such prediction methods do not provide an enhanced understanding of the principles underpinning folding and that simulated trajectories from protein structure prediction methods are inconsistent with all available experimental data, in terms of folding mechanism, kinetics, or structural data. They find that, in the general context of computational protein biophysics, current protein structure predicting programs, while now very successful at their primary role, are not an appropriate tool to investigate in vivo folding.

Pearce and Zhang (2022) are similarly skeptical, finding that AI protein folding oracles are essentially black boxes that do not help in understanding the dynamics of real-world protein folding.

As Moore et al. (2022) put it, "…[E]ven as deep learning models have achieved transformative accuracy at protein structure prediction, progress in modelling folding stability has arguably stalled."

Arguably stalled…

For those, like the writer, professionally concerned with devastating and ubiquitous disorders of protein folding like Parkinson's and Alzheimer's Diseases, *arguably stalled* is not good news.

In vivo protein folding takes place in dynamic cellular matrices under 'Clausewitzian' burdens of fog and friction exacerbated by stress and aging. The essential point is that protein folding dynamics in vivo, to be successful under those burdens, must be heavily monitored and regulated. The interested reader can run a search under the term 'protein folding chaperones' for more complete discussion (e.g., Saibil 2013).

As we will explore below in more detail, the current 'driverless car' problem faces a similar conundrum: Even for large sets of skilled and experienced drivers on transportation networks under more-or-less routine circumstances, traffic flow problems—first studied by Greenshields in the 1930s—remain essentially unsolved, for the reasons outlined by Orosz et al. (2010). Vehicle traffic flow is unlike any other flow problem, and 'phase transitions' to traffic jams and other pathologies triggered by random events are poorly understood after nearly a hundred years of intense study. See Wallace (2018) for a more complete discussion.

The essential point is that, even if AI-aided driverless cars become nearly as capable as human-driven cars, basic problems of intelligent-agent-driven-flow on complex road networks remain unsolved, although widely addressed. Of particular interest, however, will be the mixture of different 'kinds' of driverless car AI systems with human-driven vehicles. The different varieties of AI-controlled vehicles are unlikely to behave as predictably under stress as do human-controlled vehicles.

And thereupon, in a sense, hangs something of what we will say here.

We begin the formal argument by exploring recent informed speculations regarding 'breakthrough' developmentments in machine intelligence.

6.2 On 'Machine Intelligence'

Recently, Butlin et al. (2023)—some 18 others—reviewed the possibilities that artificial intelligence might become conscious, with seemingly surprising results:

> ...We survey several prominent scientific theories of consciousness, including recurrent processing theory, global workspace theory, higher order theories, predictive processing, and attention schema theory. From these theories we derive 'indicator properties' of consciousness, elucidated in computational terms that allow us to assess AI systems for these properties. We use these indicator properties to assess several recent AI systems, and we discuss how future systems might implement them. Our analysis suggests that no current AI systems are conscious, but also suggests that there are no obvious technical barriers to building AI systems which satisfy these indicators.

Wallace (2023a) extends, somewhat, the perspective:

> Consciousness in higher animals, by virtue of its 100 ms time constant, is a necessarily greatly simplified and stripped-down version of more complex multiple tunable workspace cognition/ regulation dyads like wound healing, immune function, gene expression, institutional function and the like. These more complex dynamic entities emerged through evolutionary exaptation of the inevitable information crosstalk between coresident cognitive modules. In consequence of the debrided nature of consciousness, it should not be difficult to construct a fast, single workspace 'conscious machine' that mimics the human tunable neuronal global workspace system. Tied to a 'backbrain' AI that has learned hyperrapid stereotypic pattern responses to some particular set of likely challenges, the result is an elementary 'emotional' conscious machine. A clever designer, however, may want to use available high-speed electronics to mimic the more capable multiple-workspace/workforce systems inherently less susceptible to inattentional blindness and related failings of overfocus and thrashing...

Although the term 'embodied' is indeed mentioned some 34 times in the Butlin et al. study, as a kind of recurring shibboleth, the origins and centrality of embodiment to intelligence seems curiously lacking in that work, a matter we attempt to address here and one that will lead into unexpectedly deep waters.

Shin (2023) places these matters within a 'sociotechnical' perspective,

> Algorithims are social constructs, as much as they are statistical programs. Just like any other technology, algorithms reflect and reproduce social dynamics... often intertwined with technical, cultural, and legal issues... Algorithms not only present information to users but also frame social processes and practices. They are not created from a vacuum but rather are designed and developed by humans... Algorithms are created within societies with the hope that they will align with existing social values and cultures...

6.3 Implications for HFT

It seems likely that recent developments in High Frequency Trading (HFT) for financial markets that link highly 'embodied' low-latency buy/sell algorithms to sophisticated AI 'backbrains' for time-series prediction, seemingly obviating latency problems, are now, or will soon be, approaching the limits explored in the Butlin et al. and Wallace papers. See Zhang et al. (2023) and Sezer et al. (2020) for overviews of AI in HFT and financial time series prediction.

The emergence of such 'consciousness', however, will most likely be experienced and interpreted as the punctuated occurrence of episodes of 'intense instability' that may make the flash crash of 2010 (Turiel and Aste 2022) seem tame indeed. Wallace (2005, 2012, 2022)—as do many others—explores such 'punctuation' as central to the dynamics of consciousness in a highly formal manner that we significantly expand here.

'Embodiment'—or rather its persistent lack—has, in fact, haunted AI for more than half a century (e.g., Fjelland 2020; Dreyfus 1965, 1972, 1986).

The central argument is that intelligence-as-we-know-it arose under the evolutionary selection pressure of rapid changes in embedding environmental patterns of threat and affordance (Gould 2002; Wallace 2022; 2023b). The medium for response to those embedding and enmeshing dynamics is precisely embodiment. Without hyperrapid embodiment, centering on close feedback coupling with a rich embedding world, no real 'intelligence' is possible.

Dreyfus (1965), in his opening Jeremiad against disembodied AI, places AI 'oracles' in the realm of the Alchemists:

> Alchemists were so successful in distilling quicksilver from what seemed to be dirt, that after several hundred years of fruitless effort to convert lead into gold they still refused to believe that on the chemical level one cannot transmute metals. To avoid the fate of the alchemists, it is time we asked where we stand...

A similar silo-blindness afflicted most mid-19th Century astronomers, who assumed answering the 'nebula mystery' lay in the resolution of individual stars within the masses of them that supposedly constituted such objects, via the construction of ever larger telescopes. These reached some 72 in. in diameter, without, however, resolving the supposed individual stars.

When William Huggins, in August of 1864, examined the famous 'Cat's Eye' nebula through an elementary spectroscope attached to an 8-inch telescope, the object was found to be a large mass of gas in an excited state, emitting a simple line spectrum, as opposed to the complex broad-band spectra characteristic of stars (Banish 2018). The mystery was cleanly and decisively solved, but not in the way most astronomers had assumed. (So much, it might be similarly argued, for current attempts at Chatbot AGI: bigger versions of current architectures will simply be more of the same.)

Current AI-aided HFT entities, however, facing severe and persistent Darwinian challenges, are most assuredly embodied, often on millisecond time scales. We will argue here, in a formal sense, that they are sufficiently so as to drive the emergence of very unexpected behaviors, particularly given powerful enough 'arousal'. Something much like Huggins' experiment, it seems, has already been done, although detailed results are very closely held indeed.

Dragos and Wilkins (2014) explicitly invoke evolutionary theory in the study of these matters:

> The interaction of heterogeneous actors with different time horizons and a variety of strategies produces the inefficiencies that make up a noise gradient. Ecological economics understands the market as a food web, which can be described in terms of a gain matrix defining the interdependencies between different species. At the bottom there are the basal species slaves, serfs, proletarians, free labour, consumers, savers, etc. These strata are preyed on by those further up the food chain pension funds, insurance companies, mutual funds and banks; and they in turn feed more professional financial institutions, such as hedge funds, brokers, investment banks, HFTs, etc. Each financial actor exploits the inefficiencies of the prey species and in the process produces new inefficiencies, further contributing to the 'noise gradient'. Within this complex ecology there is a gradual stabilization of predator prey relationships, but unlike an actual ecosystem, the financial system has a much higher rate of change, leading to more abrupt singular events like flash crashes evolving according to an accelerated rate of punctuated equilibria, with multiple black swans and mass extinctions (Gould and Eldredge 1977).

Dragos and Wilkins describe their work as

> ...[A]n account of the concepts of information and noise as they apply to an analysis of high-frequency trading according to heterodox economics...[W]e attempt to show how concepts from other disciplines, such as entropy, information and noise, can be useful in opening up the world of finance from its traditional economic milieu.

One way to explore this and similar arguments is to formally enlist the asymptotic limit theorems of information and control theories, as seen through the lenses of statistical physics and nonequilibrium thermodynamics. The underlying intents are both scientific insight and the construction of new statistical tools for data analysis.

6.4 Some Methodological Boilerplate

The Asymptotic Limit Theorems

Information and control theories encompass a set of constraining asymptotic limit theorems and associated lemmas (Khinchin 1957; Cover and Thomas 2006; Nair et al. 2007):

- The Shannon Coding Theorem that prescribes how to code a message in 'typical' sequences for maximum error-free transmission through a particular channel at a maximum 'channel capacity' (and the 'tuning theorem' variant in which the message formally 'transmits' the channel).
- The Source Coding Theorem that splits a set of sequences emitted by an information source into two subsets, a relatively small one that is 'meaningful'—the sequences are consonant with an underlying 'grammar' and 'syntax'—and a much

larger 'meaningless' one 'of measure zero'. Only for stationary, ergodic systems is it possible to characterize 'information' and 'information source uncertainty' in terms of a Shannon entropy-analog. For nonergodic systems, each (infinite) signal sequence has it's own scalar source uncertainty that cannot be expressed in 'Shannon entropy' form (Khinchin 1957). It is here that methods must be abducted from statistical physics, introducing as well equivalence-classes of sequences forming a groupoid, leading to groupoid symmetry-breaking phase transitions characterizing 'cognitive' phase transitions, quite unlike those of 'conventional' physical systems.
- The Rate Distortion Theorem (RDT) and an associated Rate Distortion Control Theory that, in a sense, 'tunes' the channel capacity around a maximum-permissible scalar measure of average distortion (which finesses the Friston 'free energy minimization' approach).
- The Data Rate Theorem (DRT) stating that, for stability of an inherently unstable control system, 'control information' must be provided at a rate greater than the system produces it's own 'topological information'. A version of this is, somewhat surprisingly, 'easily' derived from a time and 'free energy rate' optimization based on the Hartley-Shannon Theorem.

Further development leads to the following:

- Cognition implies choice—of a small set of actions from a larger embedding set of what is possible to the system of interest. Such choice reduces uncertainty in a formal manner, implying the existence of an information source 'dual' to the cognitive process studied, at and across the varied scales and levels of organization of the living state (Atlan and Cohen 1998; Dretske 1994; Maturana and Varela 1980).
- Real-world, real-time cognition is always inherently unstable under the impact of the 'topological information' imposed by embedding environmental 'roadways'. Invocation of the DRT then implies that *cognition must always be paired with regulation for system stability and long-term function under selection pressures.* Further, *it can be argued that the degree of pairing between cognition and regulation is, in fact, a good measure of a system's 'self awareness', a cardinal criterion in many contemporary definitions of consciousness.*
- Here, we further assume an adiabatically, piecewise, stationary (APS) 'Born-Oppenheimer' model in which the system of interest remains 'close enough' to stationary during the period(s) of interest so that the asymptotic limit theorems are effectively 'anytime algorithms' (e.g., Zilberstein 1996). Between 'pieces' for which this assumption is sufficiently true, we invoke (groupoid) symmetry-breaking and/or Fisher Zero (Dolan et al. 2001) phase transitions.
- The ergodic restriction is 'easily' lifted, at the cost of imposing methods abducted from statistical physics, with dynamics approximated via an analog to the first (and higher) order Onsager treatment of nonequilibrium thermodynamics, recognizing, however, that, due to the failure of information source microreversibility—e.g., 'eht' does not have the same probability as 'the' in English—there are no 'Onsager reciprocal relations' in high order approximations.

- For purely physical phenomena, free energy is often defined in terms of Boltzmann distribution-derived partition functions. In contrast, free energy analogs associated with the dynamics of cognitive phenomena can, and may well, be characterized by 'fat-tailed' distributions that do not have well-defined first or second moments (e.g., Derman et al. 2016; Wallace and Fricchione 2024). An explicit example will be given.

Resource Rates

Embodiment recognizes that cognitive systems—biological, institutional, mechanical, or composite—are, in reality, ultimately embedded in an environment that includes themselves. There will typically be (at least) three resource streams necessary for their function:

- The first is the rate at which elements of the system can communicate with each other, instantiated by some channel capacity $\mathcal{C}$.
- The second is the rate at which 'sensory information' is available from the embedding circumstance, say according to a channel capacity $\mathcal{H}$.
- Third, there is the rate at which 'material/materiel' resources can be provided, a rate $\mathcal{M}$. For an organism, this might be measured by the rate at which metabolic free energy is provided. Machine/institutional systems will have a different measure.

These three rates will cross-correlate: one is confronting a 3×3 (or larger) matrix $\mathbf{Z}$. An $n \times n$ matrix will have will have n scalar invariants determined by the polynomial relation

$$\begin{aligned} p(\gamma) &= \det[\mathbf{Z} - \gamma \mathbf{I}] \\ &= (-1)^n \gamma^n + (-1)^{n-1} r_1 \gamma^{n-1} + \ldots - r_{n-1}\gamma + r_n \end{aligned} \tag{6.1}$$

$\mathbf{I}$ is the appropriate identity matrix, 'det' is the determinant, and γ is a real-valued parameter. The first invariant is the matrix trace, and the last the determinant. We seek to construct a scalar resource rate measure from these invariants.

The most direct such index might be $Z = \mathcal{C} \times \mathcal{H} \times \mathcal{M}$, but our worlds are not likely so simple.

We postulate, at least in first-order, that a single scalar index $Z(r_1, \ldots, r_n)$ can be constructed that sufficiently represents information and material resource rates. This approach is analogous to, but different from, the more usual principal component analysis based on a correlation matrix. Wallace (2021) outlines the complications associated with generalization to more than a single such index.

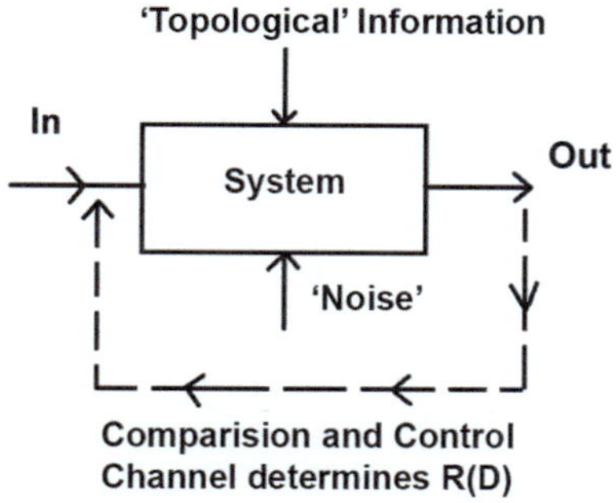

Fig. 6.1 The rate distortion control theory model. An inherently unstable system's 'topological information', according to the data rate theorem (Nair et al. 2007), is added to the effects of 'noise' to determine the minimum channel capacity $R(D)$ needed to assure transmission with average distortion D. The punctuation implied by the DRT emerges here if there is a critical maximum average distortion beyond which control fails

Rate Distortion Control Theory

A directly 'embodied' formalism emerges via a 'Rate Distortion Control Theory' based on the Rate Distortion Theorem. As described, the RDT addresses transmission under conditions of noise over an information channel. There will be, for that channel and a particular scalar measure of average distortion D between what is sent and what is received, a minimum necessary channel capacity $R(D)$.

The theorem asks, in effect, what is the 'best' channel for transmission of a message with the least possible average distortion. $R(D)$ can be defined for nonergodic information sources via a limit argument based on the ergodic decomposition of a nonergodic source into a 'sum' of ergodic sources. The RDT can be reconfigured in a control theory context, if we envision a system's topological information from the Data Rate Theorem as another form of noise, adding to the average distortion D. See Fig. 6.1.

The punctuation implied by the Data Rate Theorem (Nair et al. 2007) happens if there is a critical maximum average distortion beyond which control fails.

6.5 The Model

We provide a basic approach, applying and abducting standard formalism from statistical mechanics (Landau and Lifshitz 2007)—given a proper definition of 'temperature'—for a cognitive system.

Consider the full ensemble of high probability developmental trajectories available to the system, written as Y_j, $j = 1, 2, \ldots$. Each trajectory is associated with a Rate Distortion Function-defined minimum necessary channel capacity R_j for a particular maximum average scalar distortion D_j. Then, assuming some basic underlying probability model having distribution $\rho_R(C_R, x)$, where C_R is a parameter set, we can define a pseudoprobability for a meaningful 'message' Y_j sent into the system of Fig. 6.1 as

$$dP = \frac{\rho_R(C_R, R/g(Z))dR}{\int_0^\infty \rho_R(C_R, R/g(Z))dR} = \frac{\rho_R(C_R, R_j/g(Z))dR}{g(Z)} \tag{6.2}$$

since ρ_R has been taken a probability density that integrates to one.

Again, Y_j is a particular trajectory, while the (possibly generalized) integral is over all possible 'high probability' paths available to the system.

We implicitly impose the Shannon-McMillan Source Coding Theorem (Cover and Thomas 2006): the set of possible system trajectories can be divided in two. These are, first, a very large set of measure zero—vanishingly low probability—that is not consistent with the underlying grammar and syntax of some basic information source, and a much smaller consistent set (Khinchin 1957).

Again, R_j is the RDT channel capacity keeping average distortion less than a given limit D_j for message Y_j. $g(Z)$ is the temperature analog from above, depending on the scalar resource rate Z. $\rho(C, x)$, for physical systems, is usually taken as the Boltzmann distribution: $\rho(x) = \exp[-x]$. We suggest that, for cognitive phenomena—from the living and institutional to the machine and composite—it is necessary to move beyond analogs with physical systems. That is, it becomes necessary explore the influence of a variety of different probability distributions, including those with 'fat tails' (Derman et al. 2016; Taleb 2020), on the dynamics of cognition/regulation stability.

Again, standard methodology from statistical physics (Landau and Lifshitz 2007) identifies the denominator of Eq. (6.2) as a partition function, giving an iterated free energy-analog F as

$$\begin{aligned} \rho_F(C_F, F/g(Z)) &\equiv \int_0^\infty \rho_R(C_R, R/g(Z))dR = g(Z) \\ g(Z) &= \text{RootOf}\,(X - \rho_F(C_F, F/X)) \end{aligned} \tag{6.3}$$

where 'RootOf' means setting the expression to zero and solving for X and ρ_F is an appropriate 'free energy' distribution that must be determined case-by-case for cognitive, as opposed to simple physical, phenomena. For physical phenomena, ρ_F can often be taken as the Boltzmann distribution. Again, C_X is simply an associated parameter set.

From chemical physics (Laidler 1987), we can define a cognition rate for the system as the reaction rate analog

$$L \equiv \frac{\int_{R_0}^{\infty} \rho_R(C_R, R/g(Z))dR}{\int_0^{\infty} \rho_R(C_R, R/g(Z))dR} = \frac{1}{g(Z)} \int_{R_0}^{\infty} \rho_R(C_R, R/g(Z))dR \qquad (6.4)$$

R_0 is the minimum channel capacity needed to keep the average distortion below a critical value D_0 for the full system of Fig. 6.1.

The basic underlying probability model of Eq. (6.2)—via the partition function argument of Eq. (6.3)—determines system dynamics, but not system structure, and cannot be associated with a particular underlying network conformation, although these are related (Watts and Strogatz 1998; Barabasi and Albert 1999). More specifically, a set of markedly different networks may all be mapped on to any single given dynamic behavior pattern and, indeed, vice-versa, The same static network may display a spectrum of behaviors (Harush and Barzel 2017). We restrict focus to dynamics rather than network topology, a distinct shift from much current work.

Feynman (2000), following Bennett (1982), identifies 'information' as a form of free energy, using Bennett's clever ideal machine that turns a message directly into work. By contrast, here we are concerned here with an iterated, rather than a direct, construction.

F, as a free energy, is subject to symmetry-breaking transitions as $g(Z)$ varies (Pettini 2007). These symmetry changes, however, are not as associated with physical phase transitions as represented by standard group algebras. Such symmetry changes represent transitions from playing one 'game' to playing another. For example, a cognitive system may engage in foraging behaviors that trigger a predatory attack by another system. Then the game changes from 'foraging' to 'escape'.

Thus 'cognitive phase changes' involve shifts between equivalence classes of high probability developmental/behavioral pathways that are represented as *groupoids*, a generalization of the group concept such that a product is not necessarily defined for every possible element pair, although multiple products with multiple identity elements are defined (Brown 1987; Weinstein 1996).

Dynamics emerge via a first-order Onsager approximation akin to that of nonequilibrium thermodynamics (de Groot and Mazur 1984) in the gradient of an entropy measure S constructed from the 'iterated free energy' F of Eq. (6.3) by means of a Legendre transform that should be familiar from the Black-Scholes approximation (Landau and Lifshitz 2007; Pettini 2007).

Recall from the (first-order) Onsager approximation of nonequilibrium thermodynamics that $\partial Z/\partial t \propto \partial S/\partial Z$. The full—again, standard, first-order—development is

$$\begin{aligned}
S(Z) &\equiv -F(Z) + ZdF(Z)/dZ \\
\partial Z/\partial t &\approx dS/dZ = f(Z) \\
f(Z) &= Zd^2F/dZ^2 \\
F(Z) &= \int\int \frac{f(Z)}{Z} dZdZ + C_1 Z + C_2 \\
g(Z) &= \text{RootOf}\,(X - \rho_F(C_F, F(Z)/X))
\end{aligned} \qquad (6.5)$$

An expression for $g(Z)$ follows from the second expression of Eq. (6.3). Note that the 'diffusion coefficient' has been implicitly set equal to 1 in the second part of Eq. (6.5).

Higher order approximations are in the same realm as higher order regression models, modulo dimensional consistency. Full treatment may involve generalization of the Legendre transform via formal power series methods (Jackson et al. 2017).

There are several important points:

- Since 'RootOf' may have complex number solutions, the temperature analog $g(Z)$ enters the realm of the 'Fisher Zeros' characterizing phase transition in physical systems (e.g., Dolan et al. 2001; Fisher 1965; Ruelle 1964 Sec. 5). Such phase transitions lie at the base of the punctuated accession to consciousness in the Baars, and related, Global Workspace Models. Indeed, the 'RootOf' construction in Eq. (6.3) actually generalizes the Lambert W-function (e.g., Yi et al. 2010; Mezo and Keady 2015).
- To reiterate, information sources are not microreversible, that is, palindromes are highly improbable, for example the sequence 'ti' has far lower probability than 'it' in English. Consequently, there are no 'Onsager Reciprocal Relations' in higher dimensional systems. The imposition of groupoid symmetries on cognitive phenomena appears to follow upon this directed homotopy.
- Further, there will always be a delay in the rate of provision of Z, so that, in Eq. (6.5), for example, $f(Z) = \beta - \alpha Z(t)$—an exponential model having $Z(t) = (\beta/\alpha)(1 - \exp[-\alpha t])$—where $Z \rightarrow \beta/\alpha$ at a rate determined by α. Other dynamics are possible, such as the 'Arrhenius', $Z(t) = \beta \exp[-\alpha/t]$, with $f(Z) = (Z/\alpha)(\log(Z/\beta))^2$. However Z is constructed from the components $\mathcal{C}$, $\mathcal{H}$ and $\mathcal{M}$, so here, it is the scalar resource rate Z itself that counts.

6.6 Hazard Rates

For a simple physical system $\rho_F(C, x)$ is often the Boltzmann form of the exponential distribution

$$\rho_F(m, x) = m \exp[-mx]$$

with $m = 1$.

Cognitive entities are unlikely to be so simple. We take it possible—via selection pressure or a deliberate 'tuning' fiat—to require a particular 'hazard rate' of an embodied and regulated cognitive entity, imposing a function $Q(x)$ so that

$$\frac{\rho_F(x)}{1 - \int_0^x \rho_F(u)du} = Q(x) \tag{6.6}$$

One then solves this integral equation for the underlying probability distribution $\rho_F(x)$.

For the exponential distribution $m \exp[-mx]$ it is easy to show that $Q(x) \equiv m$, so that the exponential distribution has a uniform 'hazard rate' for signal detection/occurrence irrespective of x. Thus $Q(x)$ is the rate at which a new incoming signal will be recognized at load demand x.

We take three different cases, $Q(x) = 1$, the Boltzmann distribution, and $Q(x) = 1/(1+x) \rightarrow 0$, a detection rate declining monotonically with increasing load, and $Q(x) = x/(1+x) \rightarrow 1$, a monotonically rising rate.

Solving Eq. (6.6) for these cases gives the corresponding probability distributions as

$$\begin{aligned}
\rho_F(x) &= \exp[-x] \\
\rho_F(x) &= 1/(1+x)^2 \\
\rho_F(x) &= x\exp[-x]
\end{aligned} \tag{6.7}$$

The second distribution is 'fat-tailed' in the sense of Derman et al. (2016), and does not have finite mean or variance.

The third expression is the second-order Erlang Distribution, having finite mean and variance. It represents a decision tree through which cognitive choice—the essence of cognition—occurs as two sequential Boltzmann Distribution processes. Such compounding carries, we will show, a heavy burden. This is especially true for institutions engaged in organized conflict, representing the contrast between one-step mission command and multi-step detailed command. Russian failure in the Ukraine can be understood, in part, from this perspective.

It appears that tuning of the hazard rate—and hence the form of the basic underlying probability distribution—can be a standard response to environmental signals or other selection pressures.

6.7 Cognition Rate Examples

Base Models

cognition rate versus 'arousal' for the three specified distribution models, the Boltzmann, $1/(1+x)^2$, and the second-order Erlang, first establishing that

$$\begin{aligned}
g &= \frac{-F}{W(n, -F)} \\
g_{\pm} &= -F + \frac{1}{2} \pm \frac{\sqrt{1-4F}}{2} \\
g &= \frac{-F}{2\,W(n, -\sqrt{F}/2)}
\end{aligned} \tag{6.8}$$

where $W(n, x)$ is the Lambert W-function of order n that solves the relation

$$W(n, x)\exp[W(n, x)] = x.$$

and is real-valued only for $n = 0, -1$ with $x > -\exp[-1]$.

Note the 'Fisher Zero' phase transition conditions for the three models:

$$\begin{aligned} F &= \exp[-1] \\ F &= 1/4 \\ F &= (2\exp[-1])^2 \end{aligned} \tag{6.9}$$

Taking $R_0 = 1$ for the cognition rate relations, calculation gives

$$\begin{aligned} L &= \exp[-1/g] \\ L_+ &= \frac{g_+}{1+g_+} \\ L &= \exp[-1/g]\frac{g+1}{g} \end{aligned} \tag{6.10}$$

We take $dZ/dt = f(Z) = \beta - \alpha Z$ in Eq. (6.5) so that

$$F = \beta(Z\ln(Z) - Z) - \frac{\alpha}{Z}^2 2 + C_1 Z + C_1 \tag{6.11}$$

Fixing $\alpha = 1,\ C_1 = -1,\ C_2 = 3$, and taking $Z \equiv \beta/\alpha$, β then varies as the index of 'arousal' in, for the first two distributions, a classic Yerkes-Dodson function-vs-stress analysis (Diamond et al. 2007; Wallace and Fricchione 2024). Real-valued and imaginary components are as indicated in Fig. 6.2 for the three distribution. The two-stage Erlang Distribution, with it's motonically increasing hazard rate, however, presents some considerable contrast as there are only two very narrow real-valued stable zone spikes.

A nonzero imaginary component indicates significant instability. Nonzero imaginary components to the left, at low 'arousal' β should be interpreted as 'hallucination' while those to the right, at high arousal, indicate onset of panic or thrashing dysfunction. The intermediate zones, where the imaginary components are precisely zero, show, for the first two distributions, the classic inverted-U of Yerkes-Dodson signal transduction. Fisher Zero phase transitions indicate the changes between stability zones. Again, the two-stage Erlang Distribution presents intractable difficulties.

Interested readers can impose their own forms of the hazard rate $Q(x)$. Resulting model dynamics are not always as straightforward as Fig. 6.2, Erlang Distribution included. Beware standard computer algebra programs for the solution of integral equations.

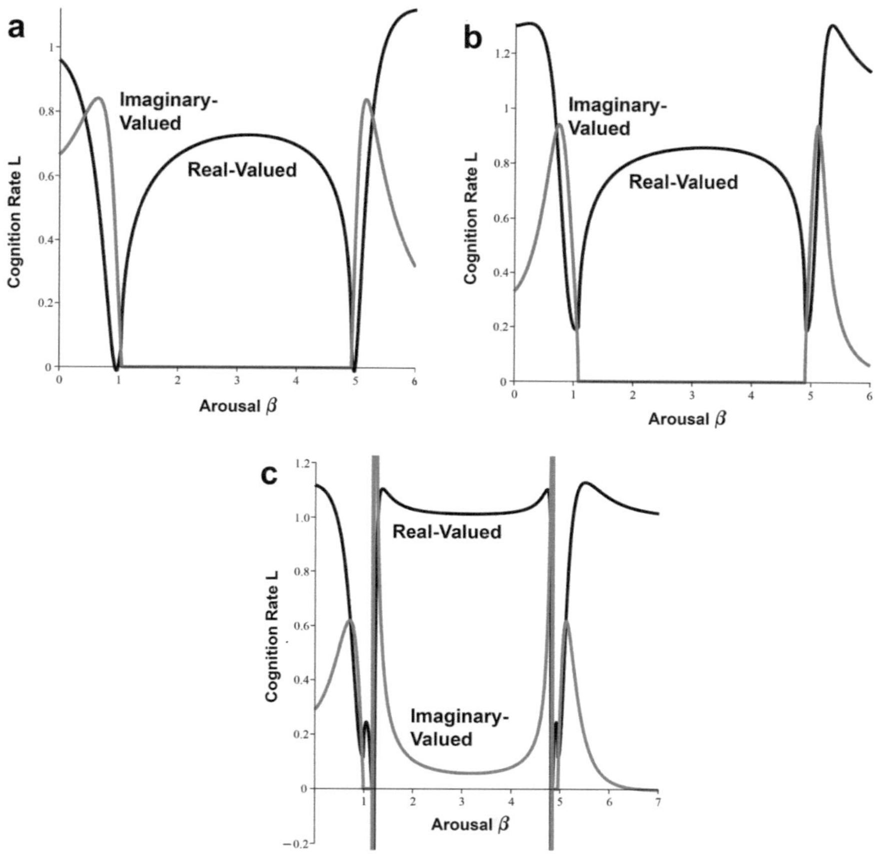

Fig. 6.2 Cognition rates versus 'arousal'. In the definition of F, $\alpha = 1$, $C_1 = -1$, $C_2 = 3$. β varies as the 'arousal' measure. **a** Exponential distribution **b** $1/(1+x)^2$ distribution. Both models are hemmed at low and high arousal by realms of 'hallucination' and 'panic' characterized by nonzero imaginary modes. The middle sections show classic inverted-U signal transduction of the Yerkes-Dodson Effect. Fisher zero transitions punctuate the phase changes. **c** The two-stage Erlang distribution, associated with a monotonically increasing hazard rate, presents difficulties, as there are only two very narrow purely real-valued stable zones

Inverted-U hazard rates, for example arising from the (fat-tailed) Cauchy Distribution—as studied in the Appendix—appear to severely limit the stability ranges of cognition rates in this model, likely to be a matter of importance across a variety of processes and venues. The essential criterion seems to be strict concavity—$d^2Q/dx^2 < 0$—over a significant range.

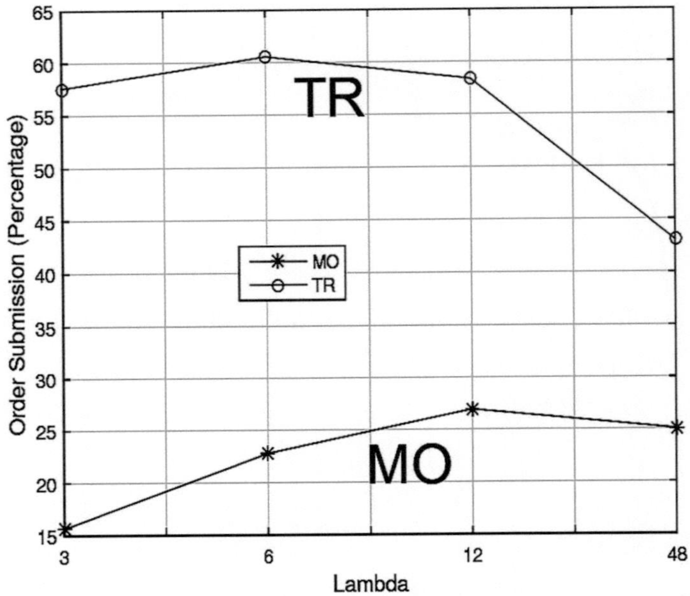

Fig. 6.3 Adapted from Fig. 4.1 of Arifovic et al. (2022). The effect of trading speed lambda of HF traders on liquidity consumption measured by market orders (MO) and the taking rate (TR)

Indeed, as the second part of the Appendix—a surprisingly concise tutorial exchange with the chatbot Perplexity AI—indicates, inverse-U relationships of various forms haunt and stalk HFT. The HFT-AI simulation model of Arifovic et al. (2022) provides an example. Figure 6.3, adapted from their Fig. 4.1, displays the effect of trading speed (lambda) of HF traders on liquidity consumption measured by market orders (MO) and the taking rate (TR).

6.8 Stochastic Excursions

Externally-Driven

What happens to cognition rate dynamics when the composite scalar 'resource rate' Z is subject to Clausewitzian fog and friction? That is, we impose a stochastic differential equation model (Protter 2006) according to the relation

$$
\begin{aligned}
dZ_t &= f(Z_t)dt + \sigma g(Z_t)dB_t \\
dZ_t &= (\beta - \alpha Z_t)dt + \sigma Z_t dB_t \qquad (6.12)
\end{aligned}
$$

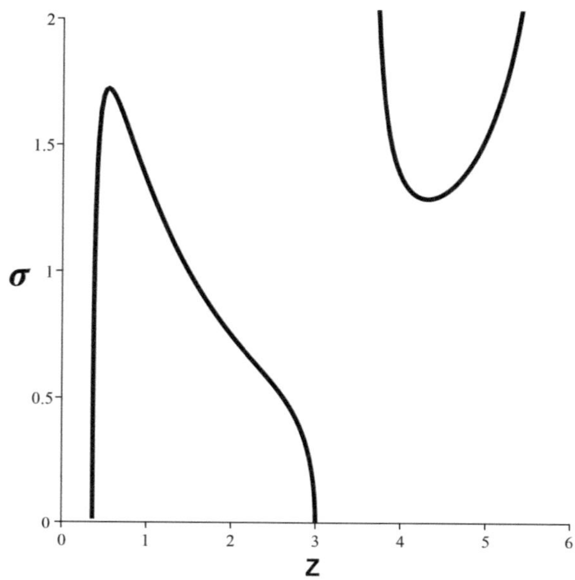

Fig. 6.4 The solution set $\{Z, \sigma\}$ for the nonequilibrium steady state cognition rate relation $< dL_t >= 0$. Increasing 'noise' σ drives the system to punctuated collapse

where dB_t represents 'colorless' Brownian stochastic variation, and for the second 'exponential' model, α instantiates 'friction', and the average $< Z_t >$ converges on β/α.

We will, in particular, seek the nonequilibium steady state solution for cognition rate, $< dL_t >= 0$, under the uniform hazard rate model. Thus g and L are given as the first expressions in Eqs.(6.8) and (6.10). We take F as in Eq.(6.11), with $\alpha = 1$, $C_1 = -1$, $C_2 = 3$, and, from Fig. 6.2a, $\beta = 3$. Then imposing the Ito Chain Rule (Protter 2006) on the expression for cognition rate L leads to an exactly solvable relation between σ and Z, producing the solution set $\{Z, \sigma\}$ of Fig. 6.4. Increasing σ drives the system to punctuated collapse, an example of noise-driven phase change, in which 'noise', in fact, becomes another form of 'temperature'.

Data Rate Theorem

A second stochastic excursion is possible via the Data Rate Theorem (DRT) that establishes the minimum necessary control information needed to stabilize an inherently unstable system (Nair et al. 2007). Details can be found in Wallace (2018, Sec. 2.1).

The essential argument starts with a linear expansion about a nonequilibrium steady state, given an n-dimensional vector of system parameters at time t. The state at time $t + 1$ is determined as

$$x_{t+1} = \mathbf{A}x_t + \mathbf{B}u_t + W_t \tag{6.13}$$

where $\mathbf{A}$, $\mathbf{B}$ are fixed $n \times n$ matrices, u_t is the n-dimensional control signal, and W_t an n-dimensional vector of Brownian white noise.

The DRT states that, under the conditions of Eq. (6.13), the minimum necessary control information rate H is given by

$$H > \log\left(|\det[\mathbf{A}^m]|\right) \equiv H_T \tag{6.14}$$

where, for $m \leq n$, $\mathbf{A}^m$ is the subcomponent of $\mathbf{A}$ having eigenvalues ≥ 1.

H_T is interpreted as the rate at which the inherently unstable system generates it's own 'topological information'. Imagine a car driven at night on a twisting, pot-holed roadway. The roadway imposes topological information at a rate determined by vehicle speed over it.

Adopting the terminology of Eq. (6.12), we extend Eq. (6.14) as

$$H(Z) > Q(Z)H_T \tag{6.15}$$

where $Q(Z)$ is monotonic increasing. The Mathematical Appendix to Wallace (2018) imposes a Black-Scholes approximation to argue that, in first order, $H(Z) \approx \kappa_1 Z + \kappa_2$. Expanding Q to similar order, so that $Q(Z) \approx \kappa_3 Z + \kappa_4$, the limit condition for stability becomes

$$K(Z) \equiv \frac{\kappa_1 Z + \kappa_2}{\kappa_3 Z + \kappa_4} > H_T \tag{6.16}$$

If $\kappa_2/\kappa_4 > H_T$ but $\kappa_1/\kappa_3 \ll H_T$, the stability condition will be violated at some sufficiently large value of Z.

K can be interpreted as a kind of temperature. If K falls below H_T without immediate 'condensation,' the system has become formally 'supercooled' and typically subject to punctuated phase transition triggered by random perturbations.

We can, in fact, apply the stochastic analysis of Fig. 6.4, calculating the nonequilibrium steady state condition $< dK_t >= 0$ under both forms of Eq. (6.12) via the Ito Chain Rule. Then, after some manipulation,

$$\begin{aligned} \sigma^2 &= \frac{f(Z)}{g(Z)^2}\left(Z + \frac{\kappa_4}{\kappa_3}\right) \\ \sigma^2 &= \frac{\beta - \alpha Z}{Z^2}\left(Z + \frac{\kappa_4}{\kappa_3}\right) \end{aligned} \tag{6.17}$$

which can be solved for $Z(\sigma)$ via complicated expressions.

Taking the underlying conditions of that figure, i.e., in Eq. (6.11) for the Boltzmann Distribution, $f(Z) \equiv \beta - \alpha Z$ with $R_0 = 1$, $\alpha = 1$, $\beta = 3$, $C_1 = -1$, $C_2 = 3$, we set $\kappa_1 = \kappa_3 = \kappa_4 = 1$, $\kappa_2 = 4$, so that, for $H_T \equiv 2$, the critical value is $Z_0 = 2$.

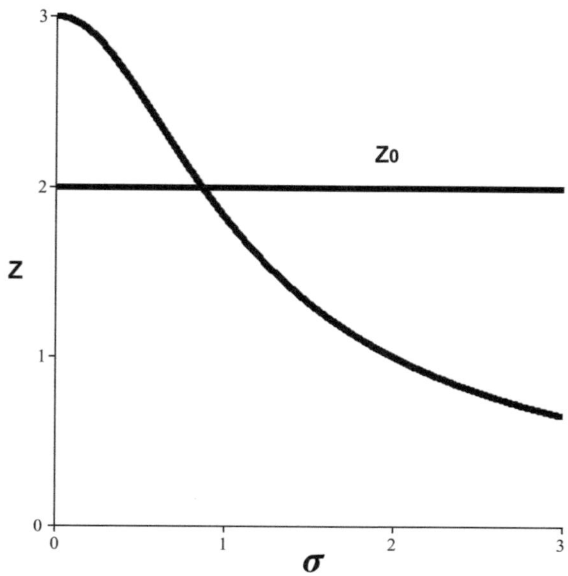

Fig. 6.5 The nonequilibrium steady state $\{Z, \sigma\}$ for $< dK_t >= 0$ from Eqs.(6.16) and (6.17). Under the conditions of Fig. 6.4, increasing 'noise' σ drives the system below the critical level DRT level H_T

Figure 6.5 shows the solution set $\{Z, \sigma\}$ for $< dK_t >= 0$ under Eq. (6.12). When $\sigma^2 \geq 3/4$, then $Z \leq 2$, and system stability fails, according to the Data Rate Theorem.

Innate and Adaptive Regulation

Instead of imposing $dZ/dt = f(Z)$ from the second line of Eq. (6.5), we take $dZ/dt \propto dS/dZ = Zd^2F/dZ^2$, letting 'noise' drive system dynamics via the stochastic differential equation

$$dZ_t = \left(Z_t d^2F/dZ^2\right) dt + \sigma Z_t dB_t \tag{6.18}$$

where, as in Eq. (6.12), dB_t represents Brownian white noise.

We assume the $1/(1+x)$ hazard rate model, so that $\rho(x) = 1/(1+x)^2$. Since this is a 'fat tailed' distribution without finite first or second orders, we impose a nonequilibrium steady state on $\ln(Z)$ via the Ito Chain Rule, assuming that, for system function under this distribution, $< d\ln(X_t) >= 0$. $F(Z)$ is, after the Ito Chain Rule algebra, then determined as

$$\begin{aligned} d^2F/dZ^2 - \sigma^2/2 &= 0 \\ F(Z) &= \frac{\sigma^2 Z^2}{4} + C_1 Z + C_2 \end{aligned} \tag{6.19}$$

We break the regulatory channel into two components with respective 'innate' and 'adaptive' capacities R and aR that, in first approximation, combine into a single totel $(1+a)R$. Some mnipultion gives

$$\begin{aligned}
\rho(F/g) &= \frac{g}{1+a} \\
L &= \int_{R_0(1+a)/g}^{\infty} \rho(u)du
\end{aligned} \tag{6.20}$$

For the $1/(1+x)$ hazard rate model,

$$\begin{aligned}
g &= -F + \frac{a}{2} + \frac{1}{2} + \frac{\sqrt{-4\,Fa + a^2 - 4\,F + 2a + 1}}{2} \\
L &= \frac{g}{g+1+a}
\end{aligned} \tag{6.21}$$

Figure 6.6a and b examine the cognition rate $L(Z, a)$ for $a = 0,\ a = 3$, setting $\sigma = 0,\ 0.6$, with $R_0 = 1,\ C_1 = -1,\ C_2 = 3$.

'Noise' σ and 'arousal' Z interact to determine cognition rate dynamics. At zero noise, larger degrees of 'adaptive' regulation broaden the domain of stable cognitive function, at the cost of maximum rate. At higher noise, however, greater adaptive regulation provides significant improvement in system stability, modulo onset of an inverse-U Yerkes-Dodson cognition pattern. As the second-order example in the Chapter Appendix will indicate, Fig. 6.18, this appears to be a ubiquitous pattern.

6.9 Time and Resource Constraints

Innate and Adaptive Regulation

We next examine a large-scale, interlinked multicomponent cognition/regulation system where there are constraints on both the resource rates Z_i and the *times to decision* T_i, so that $Z = \sum_i Z_i$ and $T = \sum_i T_i$. We seek to minimize overall distortions D_i between what is wanted and what is observed across the system-as-a-whole, using a simple Lagrange optimization, under the innate-adaptive regulation model of the previous subsection—again using the $1/(1+x)$ hazard rate model.

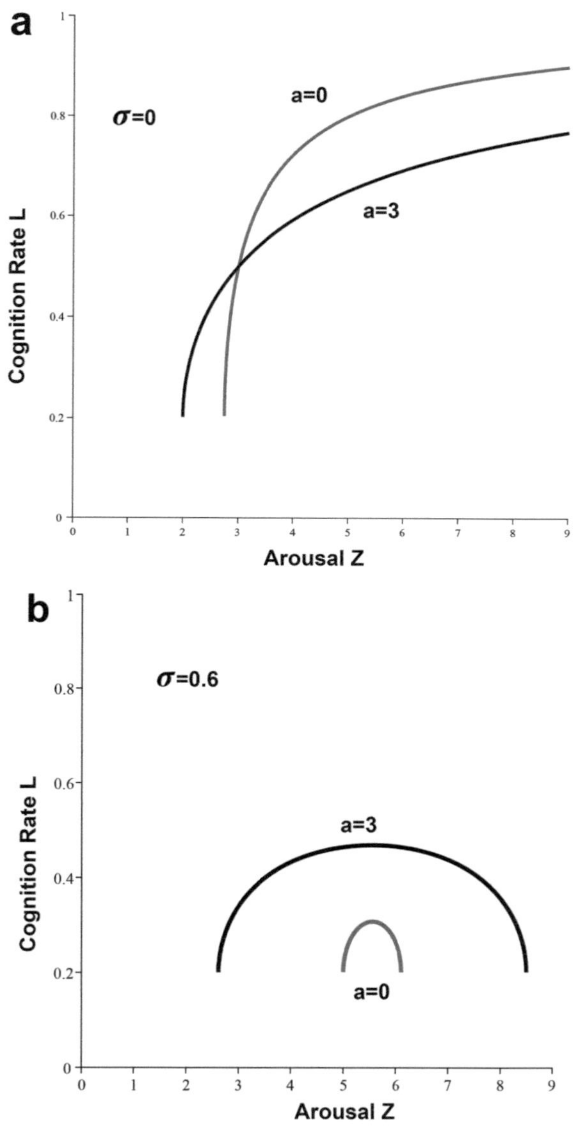

Fig. 6.6 'Innate' and 'Adaptive' regulation of cognition rate under noise for the $1/(1+x)$ hazard rate model. Degree of 'adaptiveness' a, and 'arousal' Z, interact to determine cognition rate patterns. At zero noise, **a**, larger degrees of 'adaptive' regulation broaden the domain of stable cognitive function, at the cost of maximum rate. At higher noise, **b**, however, greater adaptive regulation provides significant improvement in system stability, modulo onset of an inverse-U Yerkes-Dodson cognition pattern. See Fig. 6.18 for a similar example, in second-order

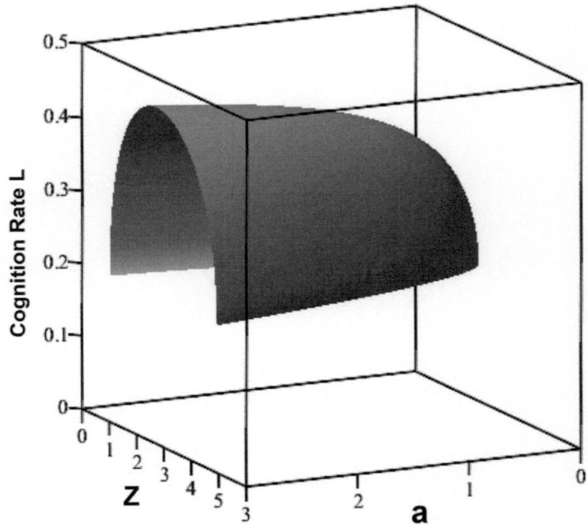

Fig. 6.7 Individual component cognition rate as a function of both Z and a, the degree to which 'adaptive' regulation is imposed above 'innate' regulation. The system collapses at $a = 0$ and is otherwise bracketed by inverse-U relations that broaden as a increases. Here, $\omega = R_0 = 1$, $C_1 = -1$, $C_2 = 3$. The 'i' designation has been dropped for convenience

We then assume Eq. (6.21), modulo a somewhat different definition of F. The—standard—Lagrangian relations are given as

$$\begin{aligned} \mathscr{L} &\equiv \sum_i D_i + \lambda\left(Z - \sum_i Z_i\right) + \mu\left(T - \sum_i T_i\right) \\ \partial\mathscr{L}/\partial Z &= \lambda,\ \partial\mathscr{L}/\partial T = \mu \\ \partial D_i/\partial Z_i &= \lambda,\ \partial D_i/\partial T_i = \mu \end{aligned} \tag{6.22}$$

In economic theory, λ and μ represent 'shadow prices' imposed on system dynamics by environmental constraints (Jin et al. 2008; Robinson 1993).

Heuristically, from Eq. (6.5), and abusing the fourth line of Eq. (6.22), we now write

$$\begin{aligned} \partial Z_i/\partial T_i &= \mu/\lambda \equiv \omega = Z_i d^2 F_i/dZ_i^2 \\ F_i &= \omega Z_i\left(\ln(Z_i) - 1\right) + C_1 Z + C_2 \end{aligned} \tag{6.23}$$

where F from Eq. (6.23) now replaces Eq. (6.19) in the calculation of cognition rates across the system. Here, we set $\lambda = \mu = R_0 = 1$, with $C_1 = -1$, $C_2 = 3$. Figure 6.7 shows the individual component cognition rate as a function of both Z and a, the degree to which 'adaptive' regulation is imposed above 'innate' regulation. In the figure, the 'i' designation has been dropped. The results are striking: the system collapses at $a = 0$ and is otherwise bracketed by inverse-U relations that broaden as a increases.

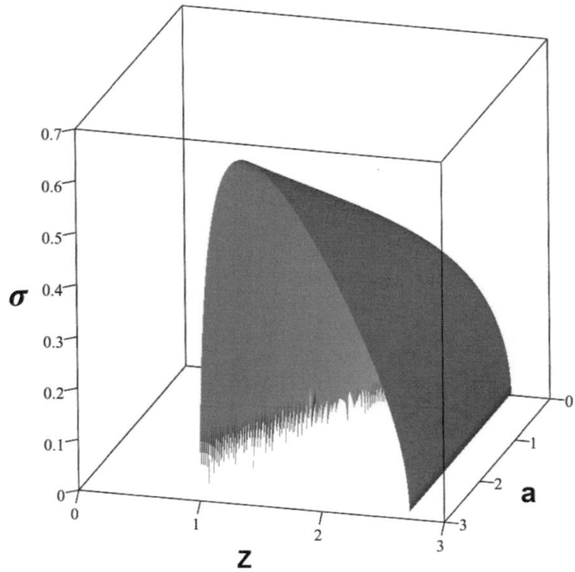

Fig. 6.8 Solution set for the nonequilibrium steady state condition $< dL_t >= 0$ under an Ito Chain rule calculation based on Eq. (6.23). Rising a, the index of 'adaptive' regulation, increases the stable realm under stochastic burden σ. At all levels, however, there is a maximum tolerable noise level. The evident tradeoff for stability under noise is a progressive decline in the resource rate Z_i

We can expand the discussion in the direction of Fig. 6.4, exploring the stability of the system under noise burden, by calculating the nonequilibrium steady state of the cognition rate in Eq. (6.21), i.e., $< dL_t >= 0$, using the Ito Chain Rule based on the stochastic differential equation

$$dZ_t = \omega dt + \sigma Z_t dB_t \tag{6.24}$$

taking parameters as in Fig. 6.7. Imposing $< dL_t >= 0$ for this model realization leads to the solution set $\{Z, a, \sigma\}$ of Fig. 6.8, again for an individual component. Increasing a, the degree of 'adaptive' regulation, expands the stable realm under 'noise' σ. In all cases, however, there is a maximum tolerable noise level. The tradeoff for stability under noise is the progressive decline in the resource rate Z_i.

It is not difficult to explore the influence of variation in the shadow price ratio ω.

Reconsidering the DRT

For the DRT model of Eqs.(6.16) and (6.17), variation in the shadow price ratio ω under time and resource constraints is of particular interest. We again impose $f(Z) = \omega$ from the Lagrangian optimization model, invoking Eq. (6.24), and calculate the nonequilibrium steady state $< dK_t >= 0$. We set $\kappa_1 = \kappa_3 = \kappa_4 = 1$, $\kappa_2 = 4$, and, again, $H_T = 2$. The result is Fig. 6.9, a three-dimensional plot analogous to Fig. 6.5. Explicitly, this is the nonequilibrium steady state solution set $\{Z, \omega, \sigma\}$ for $< dK_t >= 0$ under the constrained relation of Eq. (6.24), using the Ito Chain Rule.

Fig. 6.9 Analog to Fig. 6.5, showing the nonequilibrium steady state solution set $\{Z, \omega, \sigma\}$ for $< dK_t >= 0$ under the constrained relation of Eq. (6.24), taking $f(Z) = \omega$. The horizontal plane represents the limit condition $Z = 2$. There is no solution for negative shadow price ratio, while increasing the ratio raises the threshold for destabilization under the influence of σ

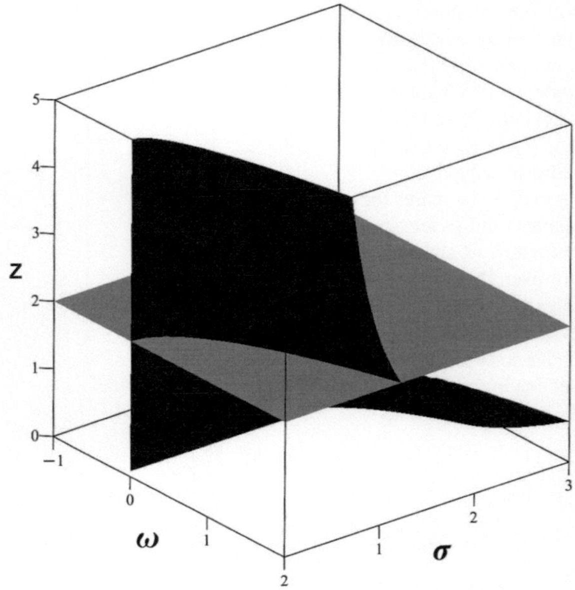

The horizontal plane is the limit condition $Z = 2$. Note that there is no stable solution at all for a negative shadow price ratio, and that increasing the ratio raises the threshold for destabilization by 'noise' σ.

6.10 Second Order Approximation

It is tedious but 'not difficult' to extend the formalism to second order. The fundamental 'Onsager' relations become

$$\begin{aligned} S &\approx -F + Z dF/dZ + \varepsilon Z^2 d^2F/dZ^2 \\ dZ/dt &\approx dS/dZ + \mu Z d^2S/dZ^2 \\ &= Z\left(\left(\frac{d^4}{dZ^4}F(Z)\right)Z^2\varepsilon\mu + 4Z\left(\left(\mu+\frac{1}{4}\right)\varepsilon+\frac{\mu}{4}\right)\left(\frac{d^3}{dZ^3}F(Z)\right)\right) \\ &\quad + Z\left(2(\mu+1)\left(\varepsilon+\frac{1}{2}\right)\left(\frac{d^2}{dZ^2}F(Z)\right)\right) \end{aligned} \tag{6.25}$$

where $\varepsilon = 1$ defines the classic second-order Legendre Transform.

Setting $dZ/dt = f(Z)$, the last expression can (again) be solved for $F(Z)$ in terms of an elaborate integral in $f(Z)/Z$.

For example, if $f(Z) = \beta - \alpha Z$, then

$$
\begin{aligned}
F(Z) &= \frac{\beta Z\,(-1+\ln(Z))}{\varepsilon+1} \\
&\quad -\frac{\alpha\, Z^2}{4\,(\mu+1)\left(\varepsilon+\frac{1}{2}\right)} \\
&\quad -\frac{\mu^2 Z^{\frac{\mu-1}{\mu}} C_1}{\mu-1} + \frac{\varepsilon^2 C_2 Z^{-\frac{1}{\varepsilon}}}{\varepsilon+1} + C_3 Z + C_4
\end{aligned}
\tag{6.26}
$$

We must further assume some particular 'fundamental probability distribution' $\rho(x)$ based on an assumed or observed hazard rate/signal detection model. Recall again that

$$
\rho(F/g) = g, \quad L = \frac{\int_{R_0}^{\infty} \rho(R/g)dR}{g} \tag{6.27}
$$

where g is the temperature analog and L is the cognition rate.

We explore a Boltzmann Distribution model, $\rho(x) = \exp[-x]$, for which

$$
g = \frac{-F}{W(-F)}, \quad L = \exp[-R_0/g] \tag{6.28}
$$

where W is again the Lambert W-function of order zero.

Here, $R_0 = 1$, and we apply Eq. (6.26), taking $\alpha = C_1 = C_2 = \varepsilon = 1$, $C_3 = -1$, $C_4 = 3$. β, the index of 'arousal', and μ, the 'second order diffusion coefficient', are allowed to vary, producing Fig. 6.10. This system generates two distinct families of inverse-U signal transduction patterns, at low and high μ, separated by a realm of instability in which complex-valued components dominate.

We can, with little difficulty, examine the effects of stochastic burden on the system of Fig. 6.10, again using the Ito Chain Rule on L from Eq. (6.28) under the second form of Eq. (6.12). The nonequilibrium steady state for $< dL_t >= 0$ is then given by the two disjunct solution sets $\{\sigma, \mu, Z\}$ shown in Fig. 6.11, taking the defining parameters as in Fig. 6.10, with $\beta = 3$ fixed across the changing values of σ and μ. Increasing σ eventually drives Z below tolerable limits.

Typically, as empirically-studied scales change—becoming significantly different in reach or altered in time domain—what had been considered second-order effects, and thus to be ignored in first-cut engineering, are progressively more intrusive. One implication of this 'law' is that, as intelligent machines (in some sense) are larger, faster, and more deeply engaged with real-time human enterprise, novel disruptive dynamics recognizably similar to Figs. 6.10 and 6.11 will emerge. A surprisingly simple, second-order, DRT-based noise-driven model leading to the classic Yerkes-Dodson effect is explored in the chapter Mathematical Appendix.

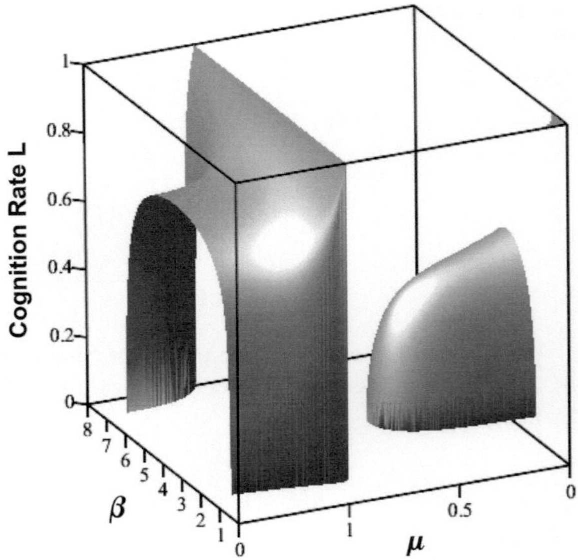

Fig. 6.10 A second-order Legendre transform model under second-order diffusion generates two distinct families of inverse-U signal transduction patterns with 'arousal' β, at low and high μ. These are separated by a realm of instability in which complex-valued components dominate

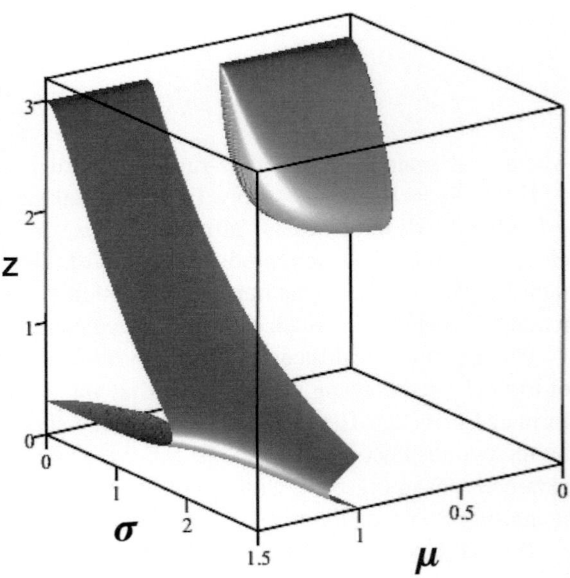

Fig. 6.11 The effects of stochastic burden on the system of Fig. 6.10, imposing the Ito Chain rule on cognition rate L from Eq. (6.28) under the second form of Eq. (6.12). The nonequilibrium steady state for $< dL_t >= 0$ has, like Fig. 6.10, two disjunct solution sets $\{\sigma, \mu, Z\}$. The parameters are as in Fig. 6.10, fixing $\beta = 3$. Sufficient σ crashes the system

Such patterns can be expected to appear in a highly punctuated manner under repeatable stressful circumstances that will often have been contemptuously dismissed as 'highly improbable' or simply 'irrelevant' under an intellectually narrow Tech Bro sales-oriented subculture.

Indeed, something of an extended case-history follows.

6.11 The 'Driverless Car' Fiasco

A peculiarity of the Western cultural pattern of 'atomistic' cognitive dynamics is the almost compulsive fixation and focus on individual—or at most, a few—'salient objects' at the expense of an 'inattentional blindness' to the embedding context (Nisbett et al. 2001; Nisbett and Miyamoto 2005). Wallace (2018) characterizes the churning driverless car fiasco in terms that clearly apply as well to current HFT policies and practices:

> An atomistic perspective on autonomous ground vehicles—seeing them as having only local dynamics in an embedding traffic stream—embodies a profound failure of insight. Traffic light strategies, road quality, the inevitably rapid-shifting 'road map space', the dynamic composition of the traffic stream, communication and machine sensory system bandwidth limits, and so on, create the synergistic context in which single vehicles operate. It is necessary to understand the dynamics of that full system, not simply the behavior of an individual vehicle atom within it. The properties of that system will be both overtly and subtly emergent—subject to sudden phase transitions into both massively and locally unstable modes—as will the responses of individual cognitive vehicles enmeshed in context, whether controlled by humans or computers. The triggering of adverse events at various scales and levels of organization by unfriendly external agents will likely become routine.

> In sum, while clever V2V/V2I [vehicle-to-vehicle, vehicle-to-infrastructure] management strategies might keep traffic streams temporarily in a 'supercooled' high-flow mode beyond well-understood critical vehicle densities, such a state is notoriously unstable, subject to both random and deliberately caused 'condensation' into large-scale frozen zones [i.e., traffic jams].

AI-driven HFT may, then, be susceptible to remarkably analogous punctuated condensations for closely analogous reasons. Wallace (2018) explores recurrent road network traffic flow patterns that may seem strangely familiar to readers. Figure 6.12, an adaptation of Fig. 1.1 of that book, displays examples of the 'fundamental diagram' of traffic flow rate versus vehicles per unit length for *individual* roadway segments in Rome, Japan, and Flanders. Respectively, the x-axes are vehicles per mile, per kilometer, and percent occupancy. There is an evident phase transition at about 40 vehicles/mile for the former two examples, and at about 10% occupancy for the latter case. Some laminar flow states may be temporarily 'supercooled' at high traffic density, but will eventually 'crystallize' out into turbulent flow traffic jams. Beyond critical loading, these road segments display hysteresis, i.e., no simple reversibility.

Cassidy et al. (2011) describe attempts to extend the idea of a fundamental diagram for an individual road segment to a full transport network:

> Macroscopic fundamental diagrams (MFDs)... relate the total time spent to the total distance traveled... It is proposed that these macrolevel relations should be observed if the data come from periods when all lanes on all links throughout the network are in either the congested or the uncontested regime...

There is some evidence for the MFD under very specific circumstances. Figure 6.13, from Geroliminis and Daganzo (2008) shows the average traffic flow versus occupancy for downtown Yokohama, where major intersections are centrally

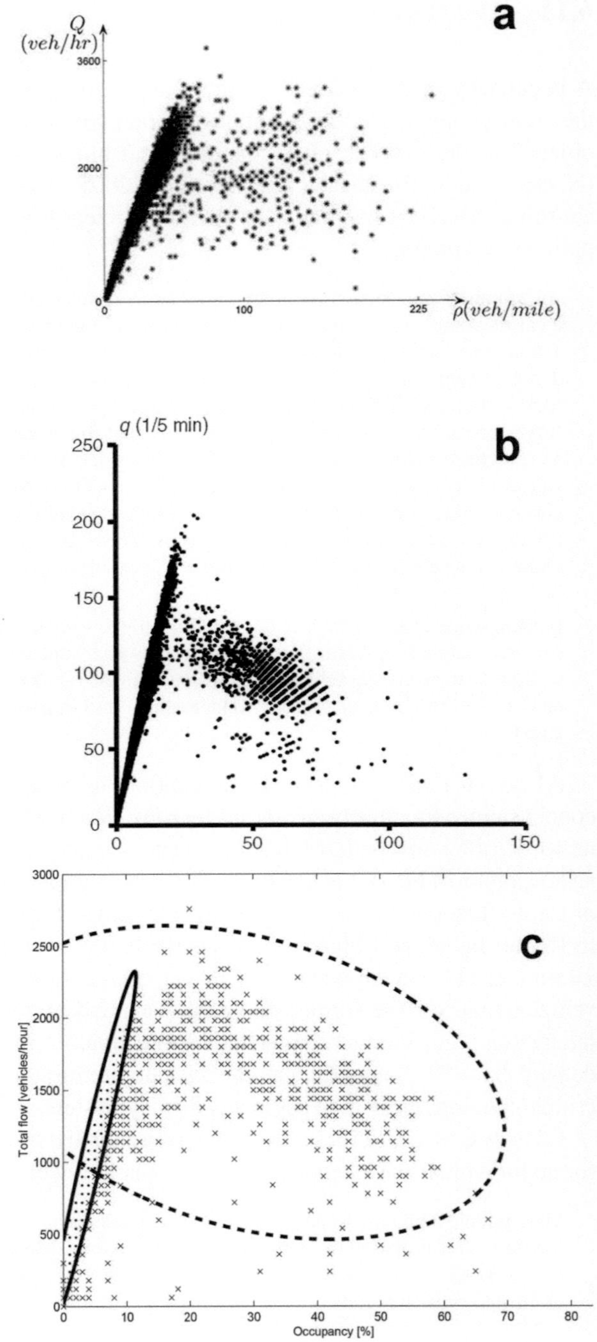

Fig. 6.12 From Wallace (2018). **a** Vehicles per hour as a function of vehicle density per mile for a street in Rome (Blandin et al. 2011). Both streamline geodesic flow and the phase transition to 'crystallized' turbulent flow at critical traffic density are evident at about 40 v/mi. Some of the states may be 'supercooled', i.e. delayed 'crystallization' in spite of high traffic density. **b** One month of data at a single point on a Japanese freeway, flow per five minutes versus vehicles per kilometer. The critical value is about 25 v/km = 39.1 v/mi (Sugiyama et al. 2008). **c** 49 Mondays on a Flanders freeway. The ellipses contain 97% of data points for the free flow and congested regimes (Maerivoet and Moor 2006). Breakdown begins at about ten percent occupancy

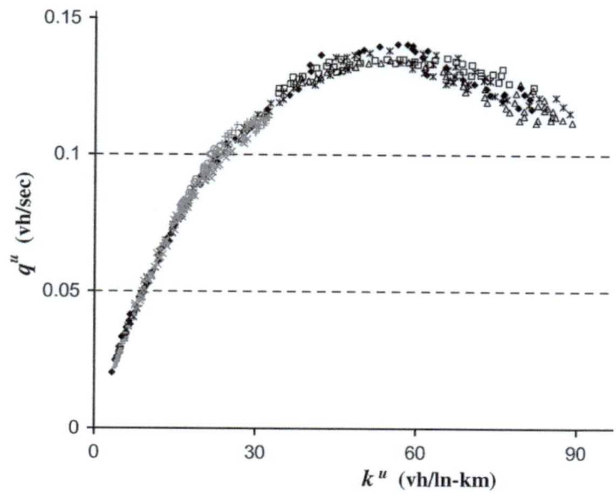

Fig. 6.13 Adapted from Geroliminis and Daganzo (2008). Average traffic flow versus occupancy (v/km) for a set of detectors in downtown Yokohama over two days. Major intersections are centrally controlled by multiphase traffic signals that respond to traffic conditions. The stability of traffic flow on the road network is thus critically dependent on the stability of the embedding active control network. The inverse-U is reversible and does not display hysteresis

controlled by multi-phase traffic signals with a cycle time that responds to traffic conditions, 110–120 sec. at night, 130–140 sec. during the day. The figure is a familiar inverse-U, *modulo the fine-tuned regulation of the multi-phase signal structure*. The network is smoothly reversible with loading, under 'proper' regulation.

However, MFD's cannot be constructed in general since congested and free-flowing sections of traffic networks will often coexist in a random manner depending on local traffic densities. Figure 6.14, adapted from Geroliminis and Sun (2011) demonstrates the limitations of the MFD approach. The figure shows the flow, in vehicles/5 min intervals, versus percent occupancy over a three-day period for the Minnesota Twin Cities freeway network that connects St. Paul and Minneapolis. While the unconstrained region of occupancy permits characterization of a smooth-flow mode, both strong hysteresis and phase transitions are evident after about eight percent occupancy, analogous to the 'nucleation' dynamics of Fig. 6.12 at higher densities. As in that figure, network flow 'fine structure' should be expected, depending on local parameters, leading to temporary 'supercooled' states. For the Twin Cities traffic network, unlike Yokohama, there is no central control that responds to traffic conditions.

One evident interpretation of Figs. 6.13 and 6.14 is that the smooth inverse-U segments of Fig. 6.2a and b—without hysteresis or nucleation—are likely to be unrealistic under current patterns of HFT-AI market regulation.

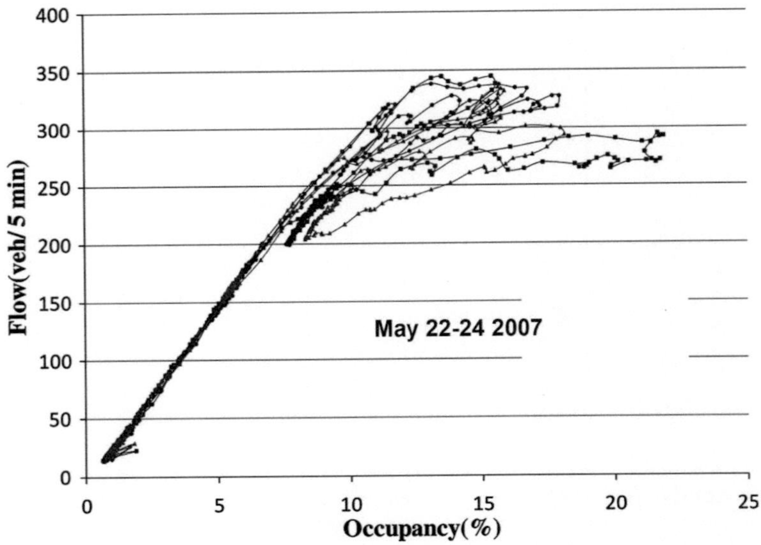

Fig. 6.14 Adapted from Geroliminis and Sun (2011). Breakdown of the macroscopic fundamental diagram for the freeway network connecting St. Paul and Minneapolis at high vehicle densities. Both nucleation and hysteresis effects are evident, showing the fine structure within the turbulent mode. As in Fig. 6.12, breakdown begins near 8–10% occupancy. Unlike Fig. 6.13, here there is no network of active control signals

6.12 Discussion

Long-persisting cognitive biological sub-systems, higher animal species, and social/institutional cognitive agents and agencies, have always closely paired cognition with regulation. Blood pressure, while necessarily responsive to physiological demand, must be kept within limits. The cognitive immune response of Atlan and Cohen (1998) must be prevented from excessive self-attack. The stream of consciousness must be held within riverbanks useful both to the individual and the social groupings within which each individual must operate. Institutional function must be held between the Scylla of inherited or learned Doctrine and the Charybdis of Law and Culture. And so on.

Current HFT agents that pair high-speed trading algorithms with AI fluctuation prediction seem to have attained levels of effective embodiment and technical sophistication that far outstrip even the most advanced realms of air combat. After all, the USA's F-35 system, while able to track and target multiple opponents simultaneously, cannot escape it's legacy as essentially an overburdened and lumbering bailout for Lockheed-Martin, all in all a true spiritual heir to the Mark 14 torpedo and the Battle of Savo Island. In real combat with near-peer adversaries, F-35 attrition rates are likely to be a singularly determining strategic factor. This is widely understood.

We must ask, then, for the global-spanning economic 'macroscopic fundamental diagram', what is the corresponding real-time regulatory structure stabilizing the domain of increasingly intelligent and fiercely competing HFT-AI agents? Again, high-speed combat aircraft are designed to be inherently unstable, permitting rapid maneuver. Control of such agents is typically done by voting among an odd number of independent—and possibly fundamentally different—computerized subsystems: throw out the outlier, average the remainder, respond. Current, heavily-funded, high-speed HFT-AI systems seem to be approaching realms of both embodiment and design that border on 'intelligence' in the sense of Dreyfus (1965, 1972, 1986) and 'consciousness' as characterized by Butlin et al. (2023). The underlying HFT-AI architecture indeed mirrors that recommended by Wallace (2023a) for development of 'machine consciousness'. Indeed, the remark on traffic flow by Orosz et al. (2010) that opens this work seems uncannily resonant.

In addition, as HFT-AI and related systems become larger and more complex, the sometimes—or even frequent—disruptive emergence of distinctly alien 'second-order' dynamics roughly analogous to those of Figs. 6.10 and 6.11 seems likely.

Thus, absent draconian regulation, increasingly sophisticated HFT-AI systems will—both *individually* and *collectively*—increasingly engage in the highly punctuated phase transitions characteristic of both intelligence and consciousness, an inevitable consequence of direct and indirect information crosstalk between cognitive submodules of larger systems (Wallace 2005; 2012; 2022). Again, the traffic flow analog seems to be found in Figs. 6.12 and 6.13.

Some of these matters have, in limited measure, entered the FinTech collective consciousness. Chapter 3 of a recent International Monetary Fund Global Financial Stability Report (IMF 2024), albeit buried behind naive fanboy gushings that 'Generative artificial intelligence and related breakthroughs have the potential to dramatically increase the efficiency of capital markets ...', raises the following cautions regarding AI implementation (IMF 2024, p.78). According to the IMF Outreach Panel interviewed, rising risks include

- Increased market speed and volatility under stress, especially if trading strategies of AI models all respond to a shock in a similar manner or shut down in response to an unforeseen event.
- More opacity and monitoring challenges, as AI spurs further migration of market-making and investment activities to hedge funds, proprietary trading firms, and other nonbank financial intermediaries and creates uncertainty about how AI models used by different investors and traders could interact.
- Increased operational risks as a result of reliance on a few key third-party AI service providers that dominate computational power and large language model services.
- Increased cyber and market manipulation risks, particularly in generating fraud and social media disinformation.
- Many of these risks are addressed by existing regulatory frameworks, but important new and unforeseen developments may arise.

Figure 6.15 below adapts Fig. 3.12 of that report:

In reply, we provide a different figure. Recall the myth of the Sorcerer's Apprentice, the 'Zauberlehrling':

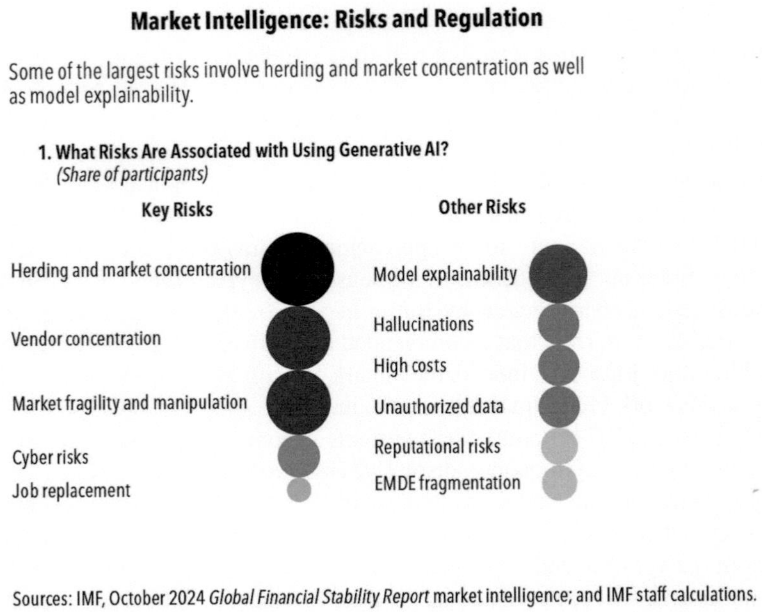

Fig. 6.15 Adapted from Fig. 3.12 of IMF (2024). Panel consensus on AI FinTech risks

Rather than inexperienced and narrow-minded Tech Bro apprentices, we badly need the services of the Hexenmeisters.

The Appendix summarizes what might be presented to an assemblage of such Hexenmeisters.

Wallace (2015) places these and similar matters within a larger milieu,

> ...[A] socioeconomic system, unlike possible biological counterparts, is a cultural artifact. There is nothing 'natural' to any particular such construct, although... large-scale dynamics remain constrained by resource availability in the context of historical trajectory and other cultural factors. Within those riverbanks, the socioeconomic system can flow according to its own inherently unstable dynamics, or it can be subject to rigorous cultural channeling... The metaphors of hunter-gatherer versus farmer are not inappropriate.

> In accordance with the Data Rate Theorem, farmed ecosystems can be inherently more productive, from a human perspective, than what can be gathered from raw nature...

> We are, very essentially, a tribe of primitive hunter-gatherers at the mercy of an unstable... monstrosity that we do not have the political will to control...

Der Zauberlehrling. F. Barth 1882. Public Domain.

Appendix

Outline for a Presentation to a Regulatory Agency

I. ***The Bottom Line***
The fundamental point is that AI-enabled High Frequency Trading is inherently and classically embodied, in contrast with bloated, hallucinating LLM chatbots. Embodiment, in a central stream of cognitive theory, is viewed as a necessary condition for intelligent behavior.

As a consequence, AI-enabled HFT machines, in the absence of draconian regulation, appear increasingly likely to display frequent, highly punctuated, 'intelligent' patterns of pathology under stress. Such behaviors challenge stabilization of market volatility and fragility and the avoidance of systemic risk and propagating shocks. An analogy with the instabilities and irregularities of traffic flow is not entirely strained.

II. ***Some Context***

1. **Perspective**
   This work is based on understanding cognition and its failure at and across various scales and levels of organization, using probability and statistical models arising from the asymptotic limit theorems of information and control theories.
   For first-order models, the theorems are extended by methods abducted from statistical mechanics and nonequilibrium thermodynamics, understanding that these only provide a gateway to more complex modeling strategies, for example including second-order effects.
   Higher-order models can be easily developed at the expense of mathematical overhead, but the explosively increasing thicket of possibilities needs to be pruned by the study of observational and experimental data.
2. **Cognition and Regulation**
   Cognition can be viewed as choice of an action, from a larger set of those possible, in response to signals from an embedding environment that includes the self. Choice reduces uncertainty, and the reduction of uncertainty implies the existence of an information source 'dual' to the process of cognition under study. The inference is direct, unambiguous, and very powerful.
   For natural systems, cognition is paired with regulation: blood pressure, the immune system, gene expression, individual consciousness, institutional cognition. Indeed, the fundamental Data Rate Theorem of control theory states that, for stability of an inherently unstable system, control information be imposed at a rate greater than the underlying 'roadway' imposes it's own 'topological information'.
   Two points:
   a. It can be argued that the degree of pairing of cognition with regulation is a good measure of a system's underlying health.
   b. The degree of a system's monitoring of it's compliance with—or deliberate deviance from—regulation may provide a sound ecological index of that system's 'self awareness,' in the classic sense.
3. **A relevant digression: 100 Milliseconds**
   The most central human mental processes operate with a time constant of about 100 milliseconds, while current, very complicated, and very expensive, fMRI brain scan phrenology has a time constant of several seconds, over tiny samples. This is insufficient for reliable scientific inference on cognition, consciousness, and their pathologies. Cutting-edge HFT-AI system time constants are well under 100 ms., and there seems to be much data available for analysis.

4. **Embodiment**
   Even given a small reactive time constant, the essential starting point for 'intelligent' machine cognition remains an embedded and reactive embodiment within, and part of, a real-time dynamic environment. This is contrary to the bloated, static chatbot LLM oracles painfully trained on materials scraped from the internet.
   Disembodied AI oracles, it can be argued, represent an inherently limited technology that may be approaching a dead end.

III ***Regulatory Concerns***

1. **How intelligent is the present crop of HFT-AI machines?**
   Compared to current AI-enabled weapons systems, HFT-AI systems appear to be the machines closest to being intelligent or conscious as these terms are currently understood across the scientific community.
   This has wide implications, not the least of which involve staged tactical or Maskirovka interventions for nefarious purposes, for example generalized spoofing attacks.
2. **Patterns of instability**
   A most fundamental driver of volatility, fragility, and systemic risk lies in the emergence of certain canonical forms of cognitive dynamics under stress. These include inverse-U and punctuated Fisher-zero phase transitions associated with groupoid symmetry-breaking. These phenomena will occur at and across critical scales and levels of organization.
   More specifically, there will be increasingly frequent individual and collective anomalous events, particularly in the face of rapid changes in degree or in rate of change—of patterns of threat or opportunity, as perceived by the machine intelligences.
   Over time, HFT-AI machines, as they evolve under selection pressure, are likely to respond ever more idiosyncratically to stress and/or its rate of change.
3. **Arrow Worms**
   From the perspectives of research in cognition and its failures, HFT-AI machines seem to provide a singular opportunity for study and analysis of proto-psychopathologies, much as experimental organisms like C. Elegans, fruit flies, and yeast serve for certain biological studies.

From policy perspectives, such proto-pathologies might seem less endearing.

Arrow worms make up the second-most common form of plankton. They are the top predator within their ecosystem, constrained only by higher-level predation. Left alone, they will consume everything in a sample, including each other. Do we really want an intelligent, indeed, conscious, Arrow Worm Trading system disconnected from regulatory or other higher-order predation?

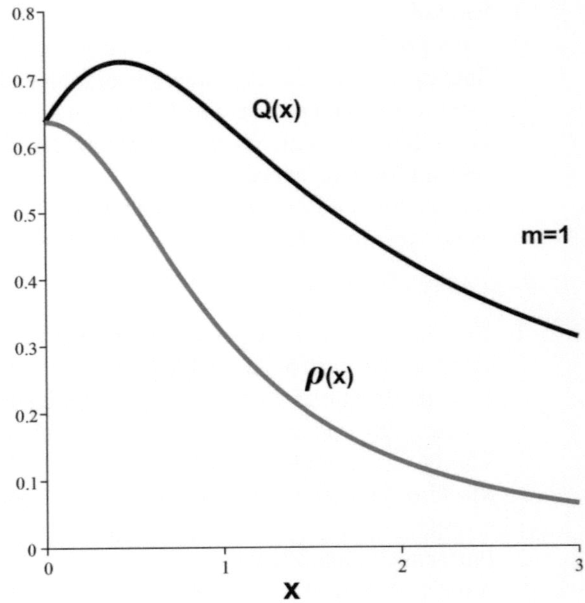

Fig. 6.16 Cauchy distribution $\rho(1, x)$ and its hazard rate $Q(1, x)$. Note the inverse-U for the hazard rate

Inverse-U Hazard Rates

We reprise the calculation leading to Fig. 6.2 for the Cauchy Distribution

$$\rho(m, x) = \frac{2}{\pi m \left(1 + \frac{x^2}{m^2}\right)}$$
$$Q(m, x) = \frac{2m}{(m^2 + x^2)\left(\pi - 2\arctan\left(\frac{x}{m}\right)\right)} \tag{6.29}$$

Figure 6.16 displays their forms for $m = 1$. Note in particular the inverse-U of the hazard rate.

The 'temperature' and cognition rates are 'easily' found to be

$$g = \frac{1 + \sqrt{-\pi^2 F^2 + 1}}{\pi m}$$
$$L = \frac{\mathrm{csgn}\left(\frac{1}{gm}\right)\pi - 2\arctan\left(\frac{R_0}{gm}\right)}{\pi} \tag{6.30}$$

where 'csgn(z)' is simply the sign of z and R_0 is a threshold value. Here, $R_0 = m = 1$.

F is as in Eq. (6.11), first, in Fig. 6.17a, setting $C_1 = -1,\ C_2 = 3$ following Fig. 6.2, and then, in Fig. 6.17b, taking $C_1 = 1,\ C_2 = 5/4$.

Figure 6.17a is even more highly punctuated than Fig. 6.2 for the same boundary conditions on F. Only a careful fitting of boundary conditions brings the two simple real-valued inverse-U peaks of cognition rate together.

Numerical exploration finds a similar pattern for cognition rates across a variety of inverse-U hazard rate forms, suggesting an underlying mathematical regularity we do not pursue here. The essential point seems to be a 'significant' range over which the hazard rate is strictly concave, i.e., $d^2Q/dx^2 < 0$.

In any event, inverse-U hazard rates appear to be particularly destabilizing for cognition rate in this model, likely a matter of some importance across a variety of critical venues.

Second-Order Yerkes-Dodson Effect

Here we adopt the second-order model, setting $\varepsilon = \mu = 1$ in Eq. (6.25), and impose a Data Rate Theorem constraint analogous to Eq. (6.16), defining $\kappa_1 = \kappa_3 = \kappa_4 = 1,\ \kappa_2 = 4$, so that $K(x)$ declines monotonically from 4 to 1 as x increases. Calculation gives

$$
\begin{aligned}
dZ/dt = {} & 3Z\left(\frac{d^2}{dZ^2}F(Z)\right) + Z^2\left(\frac{d^3}{dZ^3}F(Z)\right) \\
& + Z\left(3\frac{d^2}{dZ^2}F(Z) + 5Z\left(\frac{d^3}{dZ^3}F(Z)\right) + Z^2\left(\frac{d^4}{dZ^4}F(Z)\right)\right) \qquad (6.31)
\end{aligned}
$$

Using the base SDE $dZ_t = (dZ/dt)dt + \sigma Z_t dB_t$—where dB_t is again Brownian noise—we can then impose a nonequilibrium steady state for the DRT condition $< dK_t >= 0$ to determine the form of $F(Z)$,

$$
\begin{aligned}
F(Z) = {} & \frac{2\sigma^2 Z}{3} - \frac{\sigma^2 \ln(Z)}{3} + \frac{\sigma^2 Z^2}{12} - \frac{C_1}{2Z} - \frac{\sigma^2 \ln(Z+1)\,Z}{2} + \frac{\sigma^2}{2} \\
& -C_2\ln(Z) + \frac{\sigma^2\ln(Z+1)}{2Z} + \sigma^2 \mathrm{dilog}(Z+1) + C_3 Z + C_4 \qquad (6.32)
\end{aligned}
$$

As above, we take $C_1 = C_2 = 1,\ C_3 = -1,\ C_4 = 3$.

Imposing both the Boltzmann and $1/(1+x)^2$ probability models, so that g is given by the first two expressions of Eq. (6.8) and cognition rates by the first two of Eq. (6.10), the L versus $\{Z, \sigma\}$ graphs are shown in Fig. 6.18a and b. For both distributions, zero noise recovers the easy problem form of the Yerkes-Dodson effect, while increasing noise drives cognition into the inverse-U signal transduction mode. Note the 'hallucination' modes for both distributions at low Z. Compare with the argument leading to Fig. 6.6. The pattern is surprisingly ubiquitous.

Fig. 6.17 F is as in Eq. (6.11), with varied boundary conditions. **a** $C_1 = -1,\ C_2 = 3$. **b** $C_1 = 1,\ C_2 = 5/4$. A careful fitting of boundary conditions is needed to bring the two distinct inverse-U cognition rate peaks together

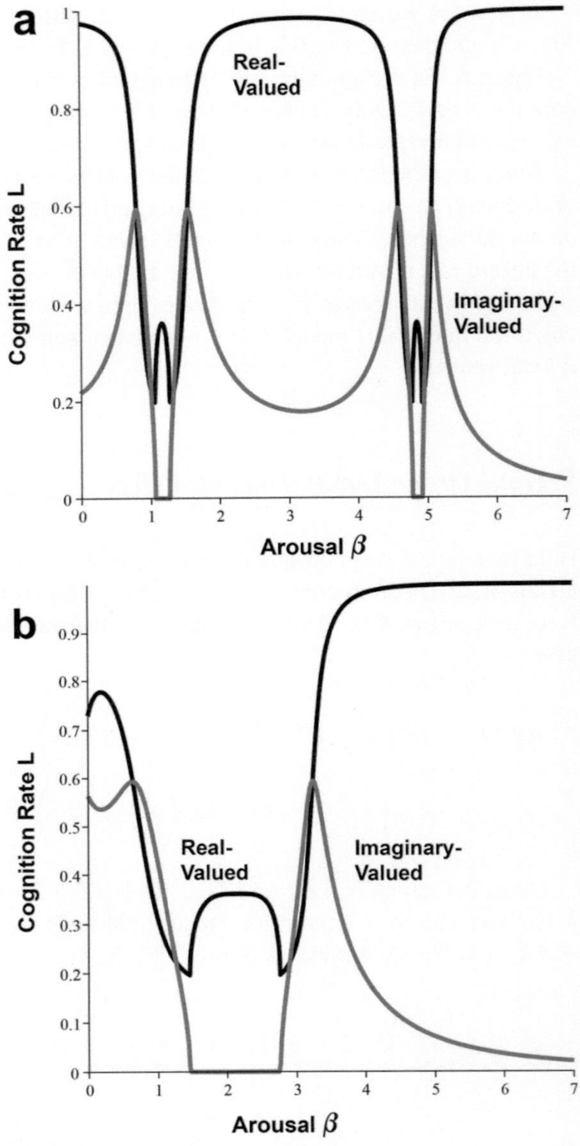

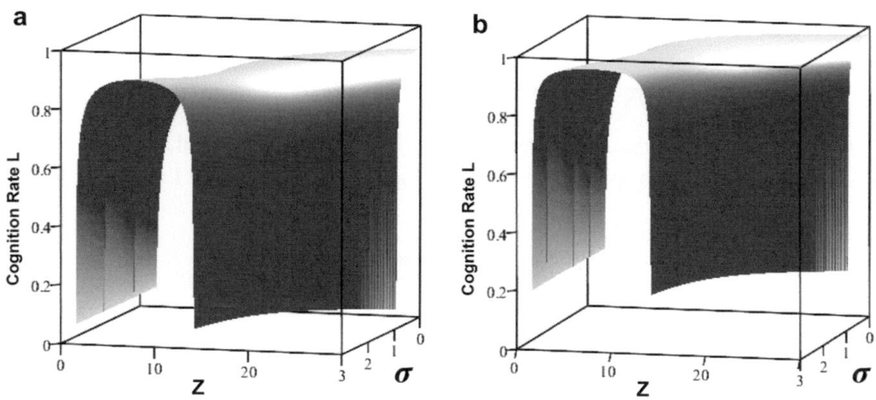

Fig. 6.18 Cognition rates as a function of 'signal' Z and 'noise' σ under a second-order model defined by the stabilization of the DRT condition. **a** The Boltzmann distribution. **b** The $1/(1+x)^2$ distribution, which is 'fat-tailed' and does not have finite mean or variance. For both, zero noise recovers the easy problem pattern of the Yerkes-Dodson effect, while increasing noise drives cognition into the inverse-U form. Note the 'hallucinations' at low Z. Compare with Fig. 6.6

Inverse-U Relationships in HFT

An exchange with Perplexity AI:

Question:
Recent empirical and theoretical studies have explored inverse-U relationships in high frequency trading. Please discuss.
Answer:
Recent studies have indeed explored inverse-U relationships in high-frequency trading (HFT), revealing interesting patterns in market dynamics and trading behavior. This phenomenon has implications for market quality, volatility, and regulatory considerations.

HFT Activity and Volatility
One notable inverse-U relationship has been observed between HFT activity and price volatility:

- At low levels of HFT activity, increasing HFT tends to reduce volatility, likely due to improved liquidity and price efficiency.
- However, beyond a certain threshold, further increases in HFT activity are associated with higher volatility.

This inverted U-shape suggests that while moderate levels of HFT can be beneficial for market stability, excessive HFT may potentially destabilize markets.

HFT Correlations Throughout the Trading Day
Another inverse-U pattern emerges when examining the correlations between HFT strategies over the course of a trading day:

- Correlations between HFT strategies increase rapidly in the first 10 min of trading.
- They continue to rise modestly until around midday.
- In the afternoon, correlations begin to decrease [8].

This pattern indicates that HFT strategies tend to be most aligned during the middle of the trading day, with more divergence in the early morning and late afternoon.

Explanations and Implications
Several factors may contribute to these inverse-U relationships:

1. Market Making versus Aggressive Trading: 'Passive HFTs engaged in market making tend to have a stabilizing effect, while more 'aggressive HFTs can potentially increase volatility when their activity becomes excessive [1].
2. Information Processing: Moderate levels of HFT can improve price discovery, but at very high levels, the speed of information processing may outpace the market's ability to absorb it efficiently.
3. Liquidity Provision: HFTs often provide liquidity, but during periods of market stress, they may withdraw liquidity simultaneously, potentially exacerbating volatility.
4. Tick Size and Adverse Selection: For stocks with smaller tick sizes, HFTs face lower adverse selection risk, which can lead to more stable trading patterns [7].

Regulatory Considerations
These findings have important implications for market regulation:

- Policymakers may need to consider optimal levels of HFT activity that balance the benefits of improved liquidity and price efficiency against the risks of excessive volatility.
- Monitoring tools that can detect when HFT activity approaches levels associated with increased volatility could be valuable for market surveillance.

Conclusion
The inverse-U relationships observed in HFT studies highlight the complex and non-linear effects of algorithmic trading on financial markets. While HFT can provide benefits in terms of liquidity and price discovery, these studies suggest that there may be an optimal level of HFT activity beyond which the marginal effects become detrimental to market quality. Continued research in this area will be crucial for developing effective policies and market structures that harness the benefits of HFT while mitigating potential risks.

Citations

1 https://www.bankofengland.co.uk/-/media/boe/files/working-paper/2012/high-frequency-trading-behavious-and-its-impact-on-market-quality-evidence-from-the-uk-equity-market.pdf.

2 https://www.sciencedirect.com/science/article/abs/pii/S0275531919309249.
3 https://www.sciencedirect.com/science/article/abs/pii/S0378437119306521.
4 https://www.sec.gov/files/dera-wp-hft-synchronizes.pdf.
5 https://www.researchgate.net/publication/332981589_High\discretionary-frequency_trading_Inverse_relationship_of_the_financial_markets.
6 https://www.academia.edu/105869255/High_frequency_trading_Inverse_relationship_of_the_financial_markets.
7 https://www.linkedin.com/pulse/decoding-cancellation-to-trade-ratio-rethinking-its-role-high-frequency-i4bac.
8 https://www.cicfconf.org/sites/default/files/paper_406.pdf.

References

Anfinsen, C., 1973, Principles that govern the folding of protein chains, Science 181:223–230.
Arifovic, J., X. He, L. Wei, 2022, Machine learning and speed in high-frequency trading, Journal of Economic Dynamics and Control 139:104438.
Atlan, H., I. Cohen, 1998, Immune information, self-organization, and meaning, International Immunology, 10:711–717.
Banish, H., 2018, The riddle of the nebulae, Sky and Telescope, August, 58–63.
Barabasi, A., R. Albert, 1999, Emergence of scaling in random networks, Science 286:509–512.
Bennett, C.H., 1982, The thermodynamics of computation, International Journal of Theoretical Physics 21:905–940.
Blandin, S. et al., 2011, A general phase transition model for vehicular traffic, SIAM Journal of Applied Mathematica 71:107–127.
Brown, R., 1987, From groups to groupoids: a brief survey, Bulletin of the London Mathematical Society, 19:113–134.
Butlin, P., et al., 2023, Consciousness in artificial intelligence: Insights from the science of consciousness, arXiv:2308.08708v3 [cs.AI] 22 Aug 2023.
Cassidy, M., K. Jang, C. Daganzo, 2011, Macroscopic fundamental diagrams for freeway networks, Transportation Research Rec 2260:8–15.
Cover, T., J. Thomas, 2006, Elements of Information Theory, Second Edition, Wiley, New York.
de Groot, S., P. Mazur, 1984, Non-Equilibrium Thermodynamics, Dover, New York.
Derman, E., M. Miller, D. Park, 2016, The Volatility Smile, Wiley, New York.
Diamond, D., A. Campbell, C. Park, J. Halonen, P. Zoladz, 2007, The Temporal Dynamics Model of Emotional Memory Processing Neural Plasticity, https://doi.org/10.1155/2007/60803.
Dolan, B., W. Janke, D. Johnston, M. Stathakopoulos, 2001, Thin Fisher zeros, Journal of Physics A 34:6211–6223.
Dragos, B., I. Wilkins, 2014, An ecological perspective on high frequency trading, Journal of Sustainable Finance and Investment 4:161–175 https://doi.org/10.1080/20430795.2014.883300.
Dretske, F., 1994, The explanatory role of information, Philosophical Transactions of the Royal Society A 349:59–70.
Dreyfus, H., 1965, Alchemy and Artificial Intelligence, Rand Report P 3244.
Dreyfus, H., 1972, What computers can't do, Harper and Row, New York.
Dreyfus, H., S. Dreyfus, 1986, Mind Over Machine, Blackwell, Basil.
Feynman, R., 2000, Lectures on Computation, Westview Press, New York.
Fisher, M., 1965 Lectures in Theoretical Physics vol. 7, University of Colorado Press, Boulder.
Fjelland, R., 2020, Why general artificial intelligence will not be realized, Humanities and Social Sciences Communications, 7:10, https://doi.org/10.1057/s41599-020-0494-4

Geroliminis, N., C. Dagazo, 2008, Existence of urban-scale macroscopic fundamental diagrams: some experimental findings, Transportation Research B 42:759–770.

Geroliminis, N., J. Sun, 2011, Properties of a well-defined macroscopic fundamental diagram for urban traffic, Transportation Research B 45:605–617.

Gould, S.J., 2002, The Structure of Evolutionary Theory, Harvard University Press, Cambridge MA.

Harush, U., B. Barzel, 2017, Dynamic patterns of information flow in complex networks, Nature Communications 8:2181

IMF, 2024, Global Financial Stability Report, October 2024, Analytical Chapters, https://www.imf.org/en/Publications/GFSR/Issues/2024/10/22/global-financial-stability-report-october-2024.

Jackson, D., A. Kempf, A. Morales, 2017, A robust generalization of the Legendre transform for QFT, Journal of Physics A 50:225201.

Jin, H., Z. Hu, X. Zhou, 2008, A convex stochastic optimization problem arising from portfolio selection, Mathematical Finance 18:171–183.

Khinchin, A., 1957, The Mathematical Foundations of Information Theory, Dover, New York.

Laidler, K., 1987, Chemical Kinetics, Third Edition, Harper and Row, New York.

Landau, L., E. Lifshitz, 2007, Statistical Physics, 3rd Ed. Part 1, Elsevier, New York.

Maerivoet, S., B. De Moor, 2006, Data quality travel time estimation and reliability, Katholieke University Leuven 06–030.

Maturana, H., F. Varela, 1980, Autopoiesis and Cognition: The Realization of the Living, Reidel, Boston.

Mezo I., G. Keady, 2015, Some physical applications of generalized Lambert functions, arXiv:1505.01555v2 [math.CA] 22 Jun 2015.

Moore, P.,W. Hendrickson, R. Henderson, A. Brunger, 2022, The protein-folding problem: Not yet solved, Science, 375(6580), 507–507.

Nair, G., F. Fagnani, S. Zampieri, R. Evans, 2007, Feedback control under data rate constraints: an overview, Proceedings of the IEEE 95, 108138.

Nisbett, R., K. Peng, C. Incheol, A. Norenzayan, 2001, Culture and systems of thought: holistic vs. analytic cognition, Psychological Review 108:291–310.

Nisbett, R., Y. Miyamoto, 2005, The influence of culture: holistic versus analytic perception, Trends in Cognitive Science 9:467–473.

Orosz, G., R. Wilaon, G. Stepan, 2010, Traffic jams: dynamics and control, Philosophical Transactions of the Royal Society A 368:4455–4479.

Pettini, M., 2007, Geometry and Topology in Hamiltonian Dynamics and Statistical Mechanics, Springer, New York.

Protter, P., 2006, Stochastic Integration and Differential Equations: A new approach, Second edition, Springer, New York.

Robinson, S., 1993, Shadow prices for measures of effectiveness II: General model, Operations Research 41:536–548.

Ruelle, D., 1964, Cluster property of the correlation functions of classical gases, Reviews of Modern Physics, April, 580–584.

Saibil, H., 2013, Chaperone machines for protein folding, unfolding and disaggregation, nature reviews molecular biology 14:630–642 https://doi.org/10.1038/nrm3658

Sezer, O., M. Gudelek, A. Ozbayoglu, 2020, Financial time series forecasting with deep learning: A systematic literature review 2005–2019, Applied Soft Computing 90:106181.

Shin, D., 2023, Algorithms, Humans, and Interactions: Designing meaningful AI experiences, Routledge, New York.

Sugiyama, Y., M. Fukui, M. Kikuchi, K. Hasebe, A. Nakayama, K. Nishinari, S. Tadaki, S. Yukawa, 2008, Traffic jams without bottlenecks: experimental evidence for the physical mechanism for the formation of a jam, New Journal of Physics, 10:033001.

Taleb, N. N. (ed.), 2020, Statistical Consequences of Fat Tails: Real world preasymptotics, epistemology, and applications, STEM Academic Press.

Turiel, J., T. Aste, 2022, Heterogeneous criticality in high frequency finance: A phase transition in flash crashes, Entropy, 24(2):257 https://doi.org/10.3390/e24020257.
Wallace, R., 2005, Consciousness: A Mathematical Treatment of the Global Neuronal Workspace Model, Springer, New York.
Wallace, R., 2012, Consciousness, crosstalk, and the mereological fallacy: an evolutionary perspective, Physics of Life Reviews, 9:426–453.
Wallace, R., 2015, An Ecosystem Approach to Economic Stabilization: Escaping the neoliberal wilderness, Routledge Advances in Heterodox Economicx, New York.
Wallace, R., 2018, Canonical Instabilities of Autonomous Vehicle Systems: The unsettling reality behind the dreams of greed, Springer, New York.
Wallace, R., 2021. How AI founders on adversarial landscapes of fog and friction, Journal of Defense Modeling and Simulation https://doi.org/10.1177/1548512920962227.
Wallace, R., 2022, Consciousness, Cognition, and Crosstalk: The evolutionary exaptation of nonergodic groupoid symmetry-breaking, Springer, New York
Wallace, R., 2023a, On 'machine consciousness', Journal of Artificial Intelligence and Consxiousness, 10:125–148, https://doi.org/10.1142/S2705078522500187.
Wallace, R., 2023b, Essays on the Extended Evolutionary Synthesis: Formalizations and expansions, Sprnger, New York.
Wallace, R., G. Fricchione, 2024, Stress-induced failure of embodied cognition: a general model, BioSystems, https://www.sciencedirect.com/science/article/abs/pii/S0303264724000789
Watts, D., S. Strogatz, 1998, Collective dynamics of small world networks, Nature 393:440–442.
Weinstein, A., 1996, Groupoids: unifying internal and external symmetry, Notices of the American Mathematical Association, 43:744–752.
Yi, S., P.W. Nelson and A.G. Ulsoy, 2010, Time-Delay Systems: Analysis and Control Using the Lambert W Function, World Scientific, New Jersey.
Zhang, X., Y. Huang, K. Xu, L. Xing, 2023, Novel modelling strategies for high-frequency stock trading data, Financial Innovation 9:39 https://doi.org/10.1186/s40854-022-00431-9
Zilberstein, S., 1996, Using anytime algorithms in intelligent systems, AI Magazine, 17:73–83.

Chapter 7
Subtleties

Abstract The dynamics of even the simplest cognitive process under the synergistic influences of fog and friction approaches the very borderlands of contemporary mathematics. One implication is that there can never be a functional 'Artificial General Intelligence' (AGI) in the current, all-encompassing, Tech Bro sense: each and every successful autonomous cognitive system must be, first, embodied in order to correct itself—to learn from a particular embedding environment, and second, carefully tuned to—trained on—that particular environment. The realms of hallucination and panic that bookend the standard Yerkes-Dodson inverse-U of aroused cognition can only be confronted by a learning dynamic that necessitates real-time embodiment. Here, we expand consideration of these matters to multidimensional resource rates and to Weber-Fechner Laws in first and second orders of approximation. Dreyfus was more than right.

7.1 Introduction

Previous chapters have imposed a Procrustean scalarization of essential resource rates and their interactions affecting cognition that is roughly analogous to using the largest component of a correlation matrix principal component analysis. Here, we will explore some extensions of that approach.

Recall the arguments leading to Eq. (1.1). The essential resource rates for a cognitive system or entity are seen as

(i) Internal channel capacity for communication within the system of interest.
(ii) Channel capacity representing the rate at which 'sensory' or 'intelligence' information is received.
(iii) The rate at which material, 'materiel', or related free energy-analogs are delivered.

The resulting probability models of cognitive dynamics under conditions of fog-and-friction fall more or less easily within the resulting one dimensional treatment. Here, some modest extensions of that approach to higher dimensions suggest the power, ubiquity, and deep subtlety of the construct.

R. Wallace, *Hallucination and Panic in Autonomous Systems*, Studies in Computational Intelligence 1217, https://doi.org/10.1007/978-3-031-95207-4_7

Indeed, Western scholars of military enterprise since the time of Clausewitz have fixated on fog and friction. Watts (2004), for example, emphasizes that

> …[F]riction has been a consistent, recurring feature of wars, not only in our own time but also as far back as the wars of Greek city states and the Persian empire. The further realization that every actor in war—from polities and nations to individual combatants and military forces—are complex adaptive systems only underscores my central argument: friction is unlikely to be eliminated from future war regardless of technological advances.

Gray (2015) extends the argument:

> The fog and friction that harass and damage strategic performance do not comprise a static set of finite challenges which can be attrited by study, let alone by machines. Every new device and mode of war carries the virus of its own technical, tactical, operational, strategic, or political negation. …The map of fog and friction is a living, dynamic one that reorganizes itself to frustrate the intrepid explorer.

The literature on fog and friction in military enterprise—and on futile schemes to eliminate them—is immense, and has been extended to the study of strategic failure in public health (Wallace 2022a).

Likewise immense is the body of formal studies of dynamic processes instantiating, or afflicted by, fog and friction. That literature is most closely associated with the analysis of stochastic differential equations (e.g., Appleby et al. 2008; Protter 2006; Mao 2007; Khasminskii 2012; Cyganowski et al. 2002), to which we refer for exact results. Here, for pedagogic purpose, we limit analysis to the kind of heuristics familiar from standard physical theory.

Beyond the inherent malaises of fog and friction lie limitations imposed by inherent models of signal perception.

7.2 Multidimensionalities

Direct Stochastic Models

Recall the Boltzmann Distribution formulation and the first-order Onsager nonequilibrium thermodynamics approximation of earlier chapters. The Boltzmann expression $\exp[-F/g] = g$ leads to expressions for the 'temperature' g and the cognition rate L as

$$
\begin{aligned}
g &= \frac{-F}{W(n, -F)} \\
L &= \exp[-H_0/g]
\end{aligned}
\tag{7.1}
$$

where H_0 is a threshold level, F is the free energy analog, and $W(n, x)$ is the Lambert W-function of order n that satisfies the relation $W(n, x)\exp[W(n, x)] = x$. It has real values only for $n = 0, -1$ and $x \geq -\exp[-1]$.

The Onsager model then defines an entropy-analog and the dynamics of its gradient in terms of a scalar independent variate $Z(t)$ as

$$\begin{aligned} S(Z) &\equiv -F(Z) + ZdF/dZ \\ dZ/dt &\propto dS/dZ = Zd^2F/dZ^2 = f(Z) \end{aligned} \quad (7.2)$$

where we implicitly set the associated diffusion coefficient equal to one.

Here, we impose an exponential form for 'friction', taking $dZ/dt = f(Z) = \beta - \alpha Z(t)$ so that $Z \to \beta/\alpha$. Simple calculation then gives

$$F(Z) = \beta\,(Z\ln(Z) - Z) - \frac{\alpha\, Z^2}{2} + C_1 Z + C_2 \quad (7.3)$$

'Fog' appears via the stochastic differential equation

$$dZ_t = (\beta - \alpha Z_t)dt + \sigma b(Z_t)dB_t \quad (7.4)$$

where dB_t is taken as Brownian white noise.

We next assume, first, in Fig. 7.1a and b, that $Z \equiv Z_1 \times Z_2 \times Z_3$ and use the Ito Chain Rule to calculate the nonequilibrium steady state condition $< dL_t >= 0$, respectively setting $b(Z) = \sigma Z,\ \sigma$.

To do this, we employ the Ito Chain Rule programs described in Cyganowski et al. (2002), developed for the computer algebra system MAPLE.

The resulting relations are algebraically far too long to display here. For simplicity, we take $Z_3 = 1$ and show $\sigma(Z_1,\ Z_2,\ 1)$ in Fig. 7.1. Here, $\alpha = 1,\ \beta = 3,\ C_1 = -1,\ C_2 = 3$, as in earlier chapters. As Appleby et al. (2008)—and others—emphasize, the form of $b(Z)$ has important implications for system dynamics. We will make much more of this below.

By contrast, for Fig. 7.1c and d, $Z \equiv \tanh(Z_1 Z_2 Z_3)$, *so that the system 'tops out' in its response to incoming signal strength*, again using $b(Z)$ as in fugues 7.1a and b, with $Z_3 = 1$. The response surfaces are relatively flattened and without bifurcation.

The Exponential Distribution $\rho(x) = m\exp[-mx]$ has a fixed uniform 'hazard rate' m defined as

$$\frac{\rho(x)}{1 - \int_0^x \rho(u)du} \quad (7.5)$$

Imposing a declining hazard rate $1/(1+x)$ generates the 'fat-tailed' distribution $1/(1+x)^2$ that does not have finite mean or variance, and for which, in contrast to Eq. (7.1),

$$\begin{aligned} g &= -F + \frac{1}{2} + \frac{\sqrt{1-4F}}{2} \\ L &= \frac{g}{g + H_0} \end{aligned} \quad (7.6)$$

Detailed calculation produces recognizable but distorted versions of Fig. 7.1. Extension to other distributions is left as an exercise.

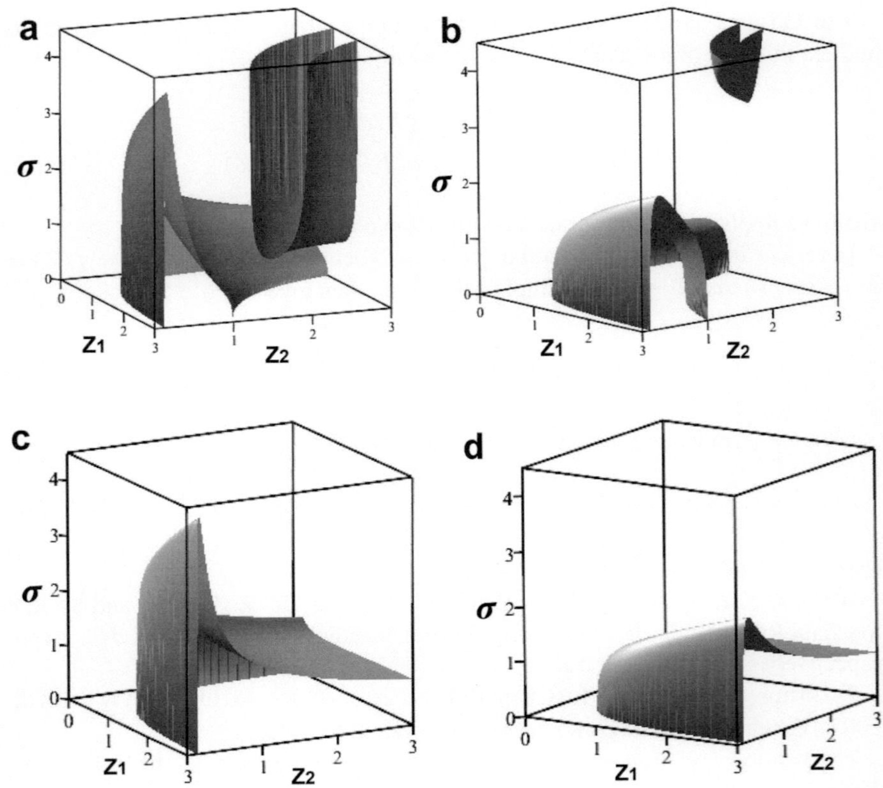

Fig. 7.1 The solution sets $\sigma(Z_1, Z_2, 1)$ for the nonequilibrium steady state $< dL_t >= 0$. **a** $Z = Z_1 Z_2 Z_3$, $b(Z) = \sigma Z$. **b** $Z = Z_1 Z_2 Z_3$, $b(Z) = \sigma$. **c** $Z = \tanh(Z_1 Z_2 Z_3)$, $b(Z) = \sigma Z$. **d** $Z = \tanh(Z_1 Z_2 Z_3)$, $b(Z) = \sigma$. For the latter two cases, the response surfaces are relatively flattened and without bifurcations

Stochastic Equipartition

Different models are, of course, possible.

Scalarization is again taken as the simple product of components, without cross-influence i.e.,

$$\begin{aligned}
Z &= \Pi_{i=1}^{n} z_i \\
\ln(Z) &= \sum_i \ln(z_i) \\
d\ln(Z)/dt &= \sum_i d\ln(z_i)/dt = \sum_i \frac{dz_i/dt}{z_i}
\end{aligned} \tag{7.7}$$

where the z_i are individual components.

Again, $dZ/dt \propto dS/dZ$, where S so that

$$\begin{aligned} dZ/dt &\propto dS/dZ = Zd^2F/dZ^2 \\ d^2F/dZ^2 &= d\ln(Z)/dt = \sum_i \frac{dz_i/dt}{z_i} \end{aligned} \tag{7.8}$$

The central problem is now the characterization of F.

Imposing the form of the z_i from Eq. (7.7), in the context of a 'dynamic equipartition', permits definition of F via the second line of Eq. (7.8).

That is, we assume

(i) Individual components z_i have been appropriately normalized.
(ii) For each such component, $dz_i/dt = \beta - \alpha z_i \; \forall i$, meaning that β and α are fixed across the system, independent of the index i.
(iii) There is an effective rapid Born-Oppenheimer-like equipartition mixing under stress so that, relative to the time frame of dynamic analysis, $\forall i, \; z_i \approx Z^{1/n}$.

Then

$$\begin{aligned} dF^2/dZ^2 &\approx n\left(\frac{\beta}{Z^{1/n}} - \alpha\right) \\ F(Z) &\approx \frac{\beta n^2 Z^{2-\frac{1}{n}}}{(-1+n)\left(2-\frac{1}{n}\right)} - \frac{n\alpha Z^2}{2} + C_1 Z + C_2 \end{aligned} \tag{7.9}$$

which differs significantly from the 'simple' result of Eq. (7.3).

Again we take the Boltzmann Distribution model, as above.

In the new expression for F of Eq. (7.9), setting $n = 3, \; \alpha = 1 \, C_1 = -10, \; C_2 = 5$—letting β again vary as the index of 'arousal'—leads to the real/imaginary-valued cognition rate Yerkes-Dodson graph of Fig. 7.2. This is, somewhat surprisingly, quite comparable to Fig. 1.3 based on Eq. (7.3).

The next step, again using the Ito Chain Rule, is to derive the nonequilibrium steady state (NSS) relation $< dL_t >= 0$ for cognition rate across the two stochastic models

$$\begin{aligned} dZ_t &= Z_t n\left(\frac{\beta}{Z_t^{1/n}} - \alpha\right) dt + \sigma Z_t dB_t \\ dZ_t &= Z_t n\left(\frac{\beta}{Z_t^{1/n}} - \alpha\right) dt + \sigma dB_t \end{aligned} \tag{7.10}$$

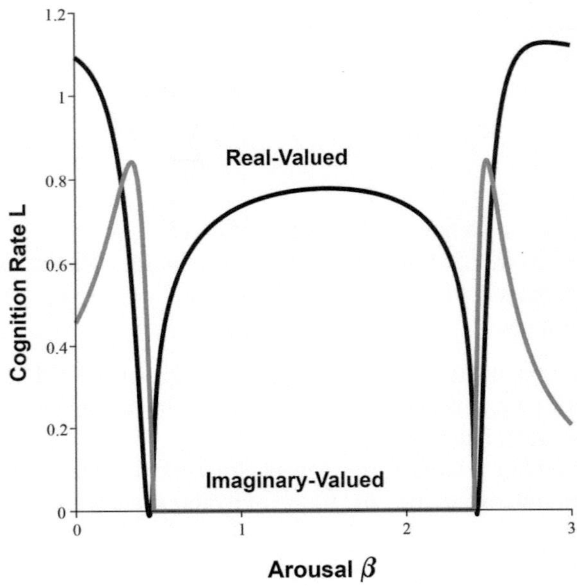

Fig. 7.2 Cognition rate calculation based on the expression for the new form of F in Eq. (7.9). The real-valued darker trace follows a Yerkes-Dodson 'arousal' pattern comparable to the 'simple' treatment of Fig. 1.3

recalling again that $dZ/dt = Zd^2F/dZ^2$, and taking dB_t as conventional Brownian white noise.

The parameter values of Fig. 7.2 give Fig. 7.3. The 'ordinary volatility' case, having the stochastic term $\sigma Z_t dB_t$, becomes unstable at a lower value of stochastic load σ.

Again, the fat-tailed $1/(1+x)^2$ probability model leading to Eq. (7.6) produces roughly analogous, if slightly distorted, results.

Less Simple Matters

We follow Wallace (2022b, Sect. 7).

There may well be more than one independent composite 'Z'-like entity irreducibly driving system dynamics. More specifically, it may be necessary to replace the scalar Z with an n-dimensional vector $\mathbf{Z}$ having independent 'orthogonal' components that, together, account for a sufficient quantity of the total variance in the rate of supply of essential resources. The dynamic equations, analogous to Eqs. (1.3)–(1.5), are of vector form,

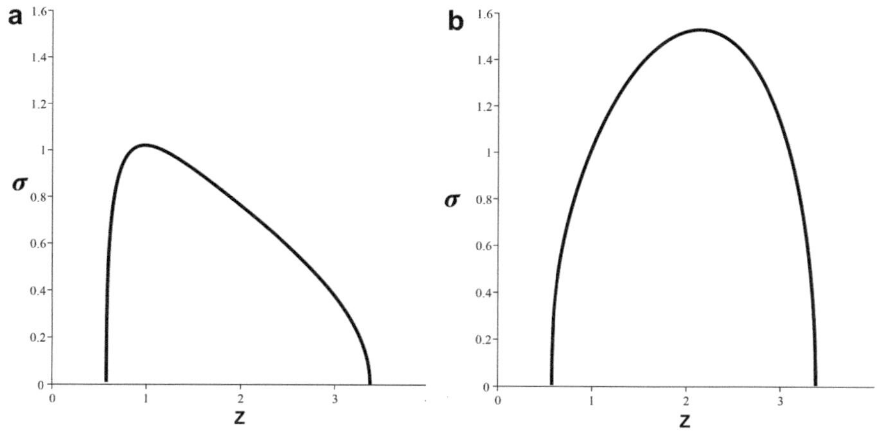

Fig. 7.3 Nonequilibrium steady state solution sets $\{Z, \sigma\}$ for $< dL_t >= 0$, using the parameter values of Fig. 7.2. **a** Ordinary volatility, $\sigma Z_t dB_t$. **b** A fully externally-driven stochastic component, σdB_t. The ordinary volatility example loses stability at a lower value of stochastic load σ

$$
\begin{aligned}
F(\mathbf{Z}) &= -\log\left(h(g(\mathbf{Z}))\right) g(\mathbf{Z}) \\
S &= -F + \mathbf{Z} \cdot \nabla_{\mathbf{Z}} F \\
\partial \mathbf{Z}/\partial t &\approx \hat{\mu} \cdot \nabla_{\mathbf{Z}} S = f(\mathbf{Z}) \\
& \quad -\nabla_{\mathbf{Z}} F + \nabla_{\mathbf{Z}}(\mathbf{Z} \cdot \nabla_{\mathbf{Z}} F) \\
&= \hat{\mu}^{-1} \cdot f(\mathbf{Z}) \equiv f^*(\mathbf{Z}) \\
\left((\partial^2 F/\partial z_i \partial z_j)\right) \cdot \mathbf{Z} &= f^*(\mathbf{Z}) \\
\left((\partial^2 F/\partial z_i \partial z_j)\right) |_{\mathbf{Z}_{nss}} \cdot \mathbf{Z_{nss}} &= \mathbf{0}
\end{aligned}
\tag{7.11}
$$

F, g, h, and S are scalar functions, $\hat{\mu}$ is an n-dimensional square matrix of diffusion coefficients. The matrix $((\partial^2 F/\partial z_i \partial z_j))$ is the obvious n-dimensional square matrix of second partial derivatives, and $f(\mathbf{Z})$ is a vector function. The last relation imposes a nonequilibrium steady state condition, $f^*(\mathbf{Z}_{nss}) = \mathbf{0}$.

For the 'simple' model, $h(g(\mathbf{Z})) \to g(\mathbf{Z})$, while, again, we assume $\mathbf{Z}(t) \to \mathbf{Z}_{nss}$.

For $n \geq 2$, this is an overdetermined system of partial differential equations. Indeed, for a general $f^*(\mathbf{Z})$, the system is inconsistent, resulting in as many as n different expressions for $F(\mathbf{Z})$, *and hence the same number of 'temperature' measures* as determined by the relation $F = -\log(g)g$ for a Boltzmann system.

This inference should not be surprising. The fifth expression in Eq. (7.11), where $f^*(\mathbf{Z}) \neq \mathbf{0}$, represents, most generally, a multi-component, cross-interacting, cross-talking, cognitive system that, if acting in real time, will almost always be far indeed from any steady state, equilibrium or nonequilibrium. Such systems will almost always be inherently unstable, requiring constant input of control information that will necessarily lag perturbation. Such a structure should not be characterizable by a single cognitive temperature-analog g. Each cognitive component, if the system is far from a steady state, should be expected to have its own g-value, in addition to structure imposed by the multidimensional nature of $\mathbf{Z}$.

Following Wallace (2022b, Sect. 7.2), we provide a two dimensional example. Suppose $\mathbf{Z} = \begin{pmatrix} x \\ y \end{pmatrix}$. Then the next-to-last expression of Eq. (7.11) becomes the partial differential equation set

$$
\begin{aligned}
x\frac{\partial^2}{\partial x^2}F(x,y) &+ y\frac{\partial^2}{\partial y \partial x}F(x,y) = f_x^*(x,y) \\
x\frac{\partial^2}{\partial y \partial x}F(x,y) &+ y\frac{\partial^2}{\partial y^2}F(x,y) = f_y^*(x,y)
\end{aligned}
\tag{7.12}
$$

The equations are inconsistent, generating two distinct formal solutions:

$$
\begin{aligned}
F^x(x,y) &= \mathscr{F}_2^x(y) + \int\int^x \frac{1}{a} f_x^*\left(a, \frac{ya}{x}\right) da\, dx \\
&+ \int \mathscr{F}_1^x\left(\frac{y}{x}\right) dx \\
F^y(x,y) &= \mathscr{F}_2^y(x) + \int\int^x \frac{1}{a} f_y^*\left(a, \frac{ya}{x}\right) da\, dy \\
&+ \mathscr{F}_1^y\left(\frac{y}{x}\right) x
\end{aligned}
\tag{7.13}
$$

$\mathscr{F}_1^q$ and $\mathscr{F}_2^q$ are boundary condition functions.

This is an example of how, in contrast to the previous one dimensional treatments, for a general n-dimensional system, there will be as many as n different 'temperature' analogs, each determining the rate of cognition of a particular modular system subcomponent. See Wallace (2022b) for further development.

These multidimensional relations are 'easily' extended to the kind of multidimensional stochastic processes described by Appleby et al. (2008). In that paper, they consider the stabilization and destabilization by a Brownian noise perturbation which preserves the equilibrium of the ordinary differential equation x (t) = f(x(t)). Under broad conditions, a function g can always be found so that the noise perturbation g(X(t)) dB(t) either stabilizes an unstable equilibrium, or destabilizes a stable equilibrium. When the equilibrium of the deterministic equation is non.hyperbolic, they show that a non.hyperbolic perturbation suffices to change the stability properties of the solution.

The essential formal point is that stochastic fog can become deeply entwined with 'simple' deterministic friction to drive unexpectedly complex patterns of failure.

7.3 Discussion

The dynamics of cognition under the often synergistic influences of fog and friction, even for the simplest of possible modalities, approaches the borderlands of contemporary mathematics. Indeed, delay stochastic differential equations, and similar matters, are particularly hot topics that we do not address here (e.g., Mao 2007; Appleby et al. 2008).

There is a singular implication: There can never be a functional 'Artificial General Intelligence' (AGI) in the current, all-encompassing, Tech Bro sense. This is because each and every successful autonomous cognitive system must be, first, embodied in order to correct itself—to learn from failure—in an embedding environment. Second, such a system must be carefully and continuously tuned to—trained on—that particular environment. The realms of hallucination and panic that bookend Fig. 7.2 in the ubiquitous Yerkes-Dodson inverse-U can only be confronted by a learning dynamic requiring real-time, real-world embodiment. Dreyfus (1965, 1972); Dreyfus and Dreyfus (1986) was right.

A venture-capital AGI-wannabe LLM that eats everything on the Internet for initial set-up cannot respond in real time to changes in a particular environment that it both inhabits and affects or afflicts.

The (highly) localized utility of the 'Arrow worm' High Frequency Trading models discussed in the previous chapter suggests that rapid-feedback embodiment can be of at least limited use in a constrained environment. Larger matters remain ambiguous: what will 'eat'—or at least regulate—artificial Arrow worms?

Indeed, the abject failure of the massive 'driverless car' initiative—a preeminently 'embodied' enterprise—suggests the inherent difficulty of ensuring the stability of cognitive cultural artifacts in the face of fog, friction, and skilled adversarial challenge (e.g., Wallace 2018, 2020, 2022a).

References

Appleby, J., X. Mao, A. Rodkina, 2008, Stabilization and destabilization of nonlinear differential equations by noise, IEEE Transactions on Automatic Control 53:683–691.

Cyganowski, S., P. Kloeden, J. Ombach, 2002, From Elementary Probability to Stochastic Differential Equations with MAPLE, Springer, New York.

Dreyfus, H., 1965, Alchemy and Artificial Intelligence, Rand Report P. 3244.

Dreyfus, H., 1972, What Computers Can't Do, Harper and Row, New York.

Dreyfus, H., S. Dreyfus, 1986, Mind Over Machine, Blackwell, Basil.

Gray, C.S., 2015, Tactical operations for strategic effect: the challenge of currency conversion, JSOU Press, MacDill Air Force Base, FL.

Khasminskii, R., 2012, Stochastic Stability of Differential Equations, Second Edition, Springer, New York.

Mao, X., 2007, Stochastic Differential Equations and Applications, Second Edition, Horwood Publishing, Chichester, UK.

Protter, P., 2006, Stochastic Integration and Differential Equations, Second Edition, Springer, New York.

Wallace, R., 2018, Canonical Instabilities of Autonomous Vehicle Systems: The unsettling reality behind the dreams of greed, Springer, New York.
Wallace, R., 2020, Cognitive Dynamics on Clausewitz Landscapes: The control and directed evolution of organized conflict, Springer, New York.
Wallace, R., 2022a, Essays on Strategy and Public Health: The systematic reconfiguration of power relations, Springer, New York.
Wallace, R., 2022b, How AI founders on adversarial landscapes of fog and friction, Journal of Defense Modeling and Simulation, 19:519–538.
Watts, B., 2004, Clausewitzian friction and future war, revised edition, McNair Paper 68. Washington, DC: Institute for National Strategic Studies, National Defense University.

Chapter 8
The Scylla and Charybdis of Intelligence: Compression Versus Complication

Abstract Data compression—separating wheat from chaff—is necessary in real-time embodied cognition, particularly on rapidly-changing 'Clausewitz landscapes' of fog, friction, and deadly adversarial intent. Here, using the asymptotic limit theorems of information and control theories, we demonstrate that, where Weber-Fechner, Stevens, Hick-Hyman, Pieron, and similar compression modes operate in the context of interfering 'noise', sufficient arousal will engender Yerkes-Dodson inverse-U response patterns for difficult problems, bracketed at low—or deliberately blunted—arousal by hallucination, and at high, by panic. With further development, the probability models explored here can be converted into statistical tools useful in the analysis of real-world, real-time data for the contrasting operational purposes of clarity or deliberate obfuscation.

> ***When, in 1961, the newly elected US president John F. Kennedy was informed about the CIA-sponsored plan to invade communist Cuba, he commissioned an appraisal from his military top brass. They concluded that the mission had a 30% chance of success—that is a 70% chance of failure. In the report that reached the president, this was rendered as 'a fair chance'. The Bay-of-Pigs invasion went ahead, and was a fiasco.***
>
> —Spiegelhalter (2024).

> ***...[I]ntelligence deficiencies played a significant role in both the failure of counterinsurgency in the first half of the 1960s and in the decision for direct military intervention in 1965. ...[R]ather than simply being a result of technical weaknesses, the lack of robust intelligence reflected wider problems of Vietnam policy, including political pressures, ideological contexts and the absence of strategic consensus...Compared to a simple case of inaccuracy or miscalculation, [a] generic form of cognitive failing proved much harder to refute or rectify by counter-evidence or counter-argument...***
>
> —Ochiai (2011).

R. Wallace, *Hallucination and Panic in Autonomous Systems*, Studies in Computational Intelligence 1217, https://doi.org/10.1007/978-3-031-95207-4_8

A wide consensus has emerged in recent years that successful policymaking and programming in conflict situations must start with an accurate understanding of local context, conflict actors, causes, and the dynamic relationships among them.

—Woocher (2011).

…[I]t is wrong for the field of intelligence studies to assume that analysts and policymakers share at least a modicum of intelligence, professional competency, and do not suffer from some mental pathology.

—Wirtz (2024).

…[A]ccurate and stable neural networks exist, however, modern algorithms do not compute them.

—Bastounis et al. (2021).

8.1 Introduction

'Sensory' or 'intelligence' data compression is essential for cognitive systems acting in real-world, real-time environments of fog, friction, and adversarial intent. Without some form of data-sorting, winnowing, or other form of compression, one is simply overwhelmed by an unredacted information stream and left paralyzed. This is, indeed, a central intent of misinformation in organized conflict.

Several such systematic compression schemes have been studied.

The Weber-Fechner Law (Kandel et al. 2013) that sensation perception is proportional to the logarithm of sensation energy has been found to approximately hold across a surprising range of modalities in human enterprise (Wikipedia 2025a; Reina et al. 2018):

- Weight perception.
- Sound intensity perception.
- Brightness perception.
- Numerical cognition.
- Dose-response chemoreception.
- Public finance in mature democracies.
- Emotional intensity.

The usual mathematical form of the Weber-Fechner Law (WF) is that the strength of the perceived 'sensation' P varies as the 'energy' of the sensory impulse S according to

$$P \approx k \ln(\frac{S}{S_0}) \tag{8.1}$$

where k is an appropriate constant and S_0 a characteristic minimum detection level.

The WF Law is most often applied to 'local' psychophysics for which results can be found with only 50 percent accuracy. So-called 'global' psychophysical observations, to higher precision, are often subsumed under the Stevens Law (Wikipedia 2025b),

$$P \approx kI^a \tag{8.2}$$

where I is an intensity measure and a an appropriate exponent. As MacKay (1963) shows, however, the Stevens Law can, in a sense, be derived from the WF Law.

From another perspective, the widely-observed Hick-Hyman Law (2028) states that the response time T_R of a cognitive agent to an incoming sensory or other data stream will be proportional to the Shannon Uncertainty of that stream. That is, more complex information takes longer to process, in direct proportion to that complexity as measured by the expression for Shannon uncertainty,

$$T_R \propto -\sum_j P_j \ln(P_j) \tag{8.3}$$

where the P_j are the *perceived* probabilities of different options.

Likewise, Pieron's Law finds response time given as (Reina et al. 2018)

$$T_R \propto I^{-b} \tag{8.4}$$

where I is the intensity of the sensory signal and b a positive constant.

Reina et al. (2018) characterize these matters as follows:

> Psychophysics, introduced in the nineteenth century by Fechner, studies the relationship between stimulus intensity and its perception in the human brain. This relationship has been explained through a set of psychophysical laws that hold in a wide spectrum of sensory domains, such as sound loudness, musical pitch, image brightness, time duration, vibrotactile frequency, weight, and numerosity. More recently, numerous studies have shown that a wide range of organisms at various levels of complexity obey these laws. For instance, Weber's law, which Fechner named after his mentor Weber, holds in humans as well as in other mammals, fish, birds, and insects. Surprisingly, also organisms without a brain can display such behaviour, for instance slime moulds and other unicellular organisms.
>
> …[F]or the first time, we show that superorganismal behaviour, such as honeybee nest-site selection, may obey the same psychophysical laws displayed by humans in sensory discriminatory tasks.

Superorganismal cognitive behavior in organized conflict, however, has long been observed to obey dynamics qualitatively similar to these psychophysics laws. Examples abound: during the Japanese attack on Pearl Harbor, as Wikipedia (2025d) describes,

> On the morning of 7 December 1941 the SCR-270 radar at the Opana Radar Site on northern Oahu detected a large number of aircraft approaching from the north. This information was conveyed to Fort Shafter's Intercept Center. The report was dismissed by Lieutenant Kermit Tyler who assumed that it was a scheduled flight of aircraft from the continental United

> States. The radar had in fact detected the first wave of Japanese Navy aircraft about to launch the attack on Pearl Harbor.

US radar subsequently—and accurately—traced Japanese aircraft in their post-attack return flights to their carriers.

In this case, essential data—complex, and, at the time, highly novel, Radar observations—were compressed to what in probability theory would be characterized as a set of measure zero. The result might well be described as a form of 'hallucination', the arbitrary assignment of meaning to a (or to the absence of a) signal.

Half a year later, however, the US Navy's institutional cognition had considerably progressed. Wirtz (2024b), in a prize-winning essay, writes

> The story of how Naval intelligence paved the way for victory at Midway is embedded in the culture of the U.S. Navy, but the impact of this narrative extends far beyond the service. Today, scholars use the events leading up to Midway to define intelligence success—an example of a specific event prediction that was accurate, timely, and actionable, creating the basis for an effective counterambush of the Imperial Japanese Navy. Yet an important element of the Midway intelligence story has been overlooked over the years: Those who received intelligence estimates understood the analysis and warnings issued and acted effectively on them.

Wirtz (2024b) emphasizes, for this success, the tight bonding of the newly-enhanced intelligence component, under Lieutenant Commander Edwin Layton, with the overall Pacific Command, under Admiral Chester W. Nimitz. This bonding represented, in a formal sense, an enhanced coupling channel capacity.

As Wirtz (2024b) put the matter, "…[O]nly nine days elapsed between Layton's forecast of coming events and the detection of Japanese carriers northwest of Midway".

This was barely sufficient time for Nimitz to assemble the needed countermeasures. The rest, as they say, is history.

In short, a sufficiently large channel capacity linking intelligence with command obviated both Pearl Harbor-style institutional hallucination and panic.

There is yet another longstanding 'psychophysics' context to such matters.

With regard to the dynamics of hallucination and panic in embodied cognitive agents, the Yerkes-Dodson Effect—studied as early as 1905 (Diamond et al. 2007; Fricchione 2023; Wallace 2024)—states that, for a simple task, 'performance' varies as an S-shaped curve with increasing 'arousal', while for a difficult task, the pattern is an inverse-U. See Fig. 8.1.

Here, we will show it possible to derive versions of Fig. 8.1 from the Weber-Fechner, Stevens, Hick-Hyman, and Pieron Laws—and analogous de-facto information compression mechanisms—using probability models based on the asymptotic limit theorems of information and control theories. These are, in turn, interpreted through the lenses of first and second order models abducted from statistical physics and the Onsager approximation to nonequilibrium thermodynamics. While, at least to the author, surprisingly straightforward, this effort is not without subtlety.

With some further—considerable—investment of resources, the probability models explored here can be developed into robust statistical tools for application to a

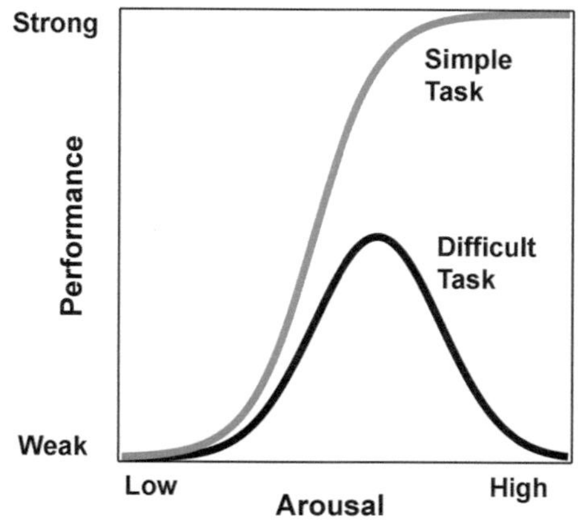

Fig. 8.1 The Yerkes-Dodson effect for simple and difficult tasks: S-shaped and inverse-U with rising 'arousal'

broad spectrum of embodied cognitive phenomena at and across various scales and levels of organization. Such tools would be particularly useful in systems where it is necessary to condense incoming 'sensory' and/or 'intelligence' information in real time. Possible applications to embodied entities and processes range from the cellular through the neural, individual, machine, institutional, and their many composites.

As Wirtz 2024b indicates, those engaged in the planning, direction, or study of organized conflict will be sadly familiar with the general dynamics studied here.

We begin with some necessary methodological boilerplate.

8.2 Rate Distortion Control Theory

French et al. (1999) modeled the Yerkes-Dodson Effect using a neural network formalism, interpreting the underlying mechanism in terms of a necessary real-time data compression that eventually fails as task complexity increases. Data compression implies matters central to information theory, for which there is a formal context, i.e., the Rate Distortion Theorem (Cover and Thomas 2006).

Information theory has three classic asymptotic limit theorems (Cover and Thomas 2006, Khinchin 1957):

- The Shannon Coding Theorem. For a stationary transmission channel, a message recoded as 'typical' with respect to the probabilities of that channel can be transmitted without error at a rate C characteristic of the channel, it's capacity. Vice versa, it is possible to argue for a tuning theorem variant in which a transmitting channel is tuned to be made typical with respect to the message being sent, so that, formally, the channel is 'transmitted by the message' at an appropriate dual channel capacity.
- The Shannon-McMillan or Source Coding Theorem. Messages transmitted by an information source along a stationary channel can be divided into sets, a very

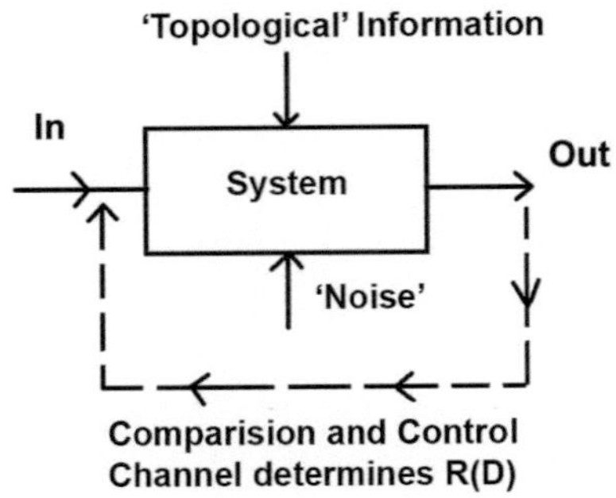

Fig. 8.2 A control system reinterpreted through the rate distortion theorem. An unstable system's 'topological information' (Nair et al. 2007) is added to 'noise' to determine the minimum channel capacity $R(D)$ needed for transmission with average distortion D. The punctuation implied by the DRT emerges from the RDT if there is a critical maximum average distortion. Some form of real-time 'embodiment' is inherent to the existence of a feedback loop

small one congruent with a characteristic grammar and syntax, and a much larger one of vanishingly small probability not so congruent. For stationary (in time), ergodic sources, where long-time averages are cross-sectional averages, the splitting criterion dividing the two sets is given by the classic Shannon uncertainty. For nonergodic sources, which are likely to predominate in biological and ecological circumstances, matters require a 'splitting criterion' to be associated with each individual high-probability message (Khinchin 1957).

- The Rate Distortion Theorem (RDT). This involves message transmission under conditions of noise for a given information channel. There will be, for that channel, assuming a particular scalar measure of average distortion, D, between what is sent and what is received, a minimum necessary channel capacity $R(D)$. The theorem asks what is the 'best' channel for transmission of a message with the least possible average distortion. $R(D)$ can be defined for nonergodic information sources via a limit argument based on the ergodic decomposition of a nonergodic source into a 'sum' of ergodic sources.

The RDT can be reconfigured in a control theory context, if we envision a system's topological information from the DRT as 'simply' another form of noise, adding to the average distortion D. See Fig. 8.2. The punctuation implied by the DRT (Nair et al. 2007) emerges from this model if there is a critical maximum average distortion that characterizes the system. Other systems may degrade more gracefully, or, as we will show below, have even more complicated patterns of punctuation that, in effect, generalize the Data Rate Theorem.

We will expand perspectives on the dynamics of cognition/regulation dyads across rates of arousal, across the possibly manifold set of basic underlying probability distributions that characterize such dyads at different scales, levels of organization, and indeed, across various intrinsic patterns of arousal.

Inherent to Fig. 8.2, of course, is a real-time 'embodiment' instantiated by the Comparison and Control Channel feedback loop.

8.3 Scalarizing Essential Resource Rates

The RDT holds that, under a given scalar measure of average distortion D between a sequence of signals that has been sent and the sequence that has actually been received in message transmission—measuring the difference between what was ordered and what was observed—there is a minimum necessary channel capacity $R(D)$ determined by the rate at which a set of essential resources, indexed by some scalar measure Z, can be provided to the system 'sending the message' in the presence of noise and an opposing rate of topological information characteristic of the inherent instability of the control system under study.

The Rate Distortion Function (RDF) $R(D)$ is necessarily convex in D, so that $d^2R/dD^2 \geq 0$ (Cover and Thomas 2006). For a nonergodic process in which the cross-sectional mean is not the same as the time-series mean, the RDF can still be defined as the average across the RDF's of the ergodic components of that process, and is thus convex in D.

The relations between the minimum necessary channel capacity $R(D)$ and the rates of any essential resources will usually be quite subtle.

A 'simple' scalar resource rate index—say Z—which we adopt as a first approximation, must be composed of a minimum of three interacting components:

- The rate at which subcomponents of the system of interest can communicate with each other, defined by a channel capacity $\mathcal{C}$.
- The rate at which 'sensory' information is available from the embedding environment, associated with a channel capacity $\mathcal{H}$.
- The rate $\mathcal{M}$ at which 'metabolic' or other free energy 'materiel' resources can be provided to a subsystem of a full entity, organism, institution, machine, and so on.

These rates must be compounded into an appropriate scalar measure

$$Z(\mathcal{C}, \mathcal{H}, \mathcal{M}).$$

Most simply, this might perhaps be taken as their product, the sum of their logs, or some other such.

Following Wallace and Fricchione 2024, 'Z' will most likely be a 3 by 3 matrix, including necessary interaction crossterms. An n-dimensional matrix $\mathbf{Z}$ has n scalar invariants r_i under appropriate transformations that are determined by the characteristic equation

$$\det[\mathbf{Z} - \gamma\mathbf{I}] = \sum_{i=1}^{n}(-1)^{i-1}\gamma^{i-1}r_i \tag{8.5}$$

$\mathbf{I}$ is the n-dimensional identity matrix, det the determinant, and γ a real-valued parameter. r_1 is the matrix trace and r_n the matrix determinant. Given these n scalar invariants, it will often be possible to construct Z as a scalar function of them, analogous to the principal component analysis of a correlation matrix.

Wallace (2021a) provides an example in which two such indices are minimally necessary, a matter requiring sophisticated Lie Group methods.

8.4 The Fundamental Model

Feynman (2000) and Bennett (1982) argue that information is a form of free energy, not an 'entropy', in spite of information theory's Shannon uncertainty taking the same mathematical form as entropy for a simple—indeed, simplistic—ergodic system. Feynman (2000) illustrates this equivalence using a simple ideal machine to convert information from a message into useful work.

We next apply something much like the standard formalism of statistical mechanics—given a proper definition of 'temperature'—for a cognitive, as opposed to a 'simply' physical, system.

We consider the full ensemble of high probability developmental trajectories available to the system, writing these as Y_j, $j = 1, 2, \ldots$. Each trajectory is associated with a Rate Distortion Function-defined minimum necessary channel capacity R_j for a particular maximum average scalar distortion D_j. Then, assuming some basic underlying probability model having distribution $\rho(c, x)$, where c is a parameter set, we can define a pseudoprobability for a meaningful 'message' Y_j sent into the system of Fig. 8.2 as

$$dP = \frac{\rho(c, R/g(Z))dR}{\int_0^\infty \rho(c, R/g(Z))dR} = \frac{\rho(c, R_j/g(Z))dR}{g(Z)} \tag{8.6}$$

since ρ is a probability density that integrates to unit value.

Again, Y_j is a particular trajectory, while the (possibly generalized) integral is over all possible 'high probability' paths available to the system.

We implicitly impose the Shannon-McMillan Source Coding Theorem so that the overall set of possible system trajectories can be divided in two distinct equivalence classes defining a 'fundamental groupoid' whose 'symmetry-breaking' imposes the essential two-fold structure. These are, first, a very large set of measure zero—vanishingly low probability—that is not consistent with the underlying grammar and syntax of some basic information source, and a much smaller consistent set (Khinchin 1957).

Again, R_j is the RDT channel capacity keeping average distortion less than a given limit D_j for message Y_j. $g(Z)$ is a yet to be determined temperature analog depending on the scalar resource rate Z. $\rho(c, x)$, for physical systems, is usually taken as the Boltzmann distribution: $\rho(x) = \exp[-x]$. We suggest that, for cognitive phenomena—from the living and institutional to the machine and composite—it is necessary to move beyond analogs with physical system analogs. That is, it becomes necessary explore the influence of a variety of different probability distributions, including those with 'fat tails' (Derman et al. 2016; Taleb 2020), on the dynamics of cognition/regulation stability.

Standard methodology from statistical physics (Landau 2007) identifies the denominator of Eq. (8.6) as a *partition function*. This allows definition of an *iterated* free energy-analog F as

$$\rho(c, F/g(Z)) \equiv \int_0^\infty \rho(c, R/g(Z))dR = g(Z) \tag{8.7}$$

Again adapting a standard argument, now from chemical physics (Laidler 1987), we can define a cognition rate for the system as a reaction rate analog

$$L \equiv \frac{\int_{R_0}^\infty \rho(c, R/g(Z))dR}{\int_0^\infty \rho(c, R/g(Z))dR} = \frac{1}{g(Z)} \int_{R_0}^\infty \rho(c, R/g(Z))dR \tag{8.8}$$

R_0 is the minimum channel capacity needed to keep the average distortion below a critical value D_0 for the full system of Fig. 8.2.

The basic underlying probability model of Eq. (8.6)—via Eq. (8.8)—determines system dynamics, but not system structure, and cannot be associated with a particular underlying network form, although they are related (Watts and Strogatz 1998; Barabasi and Albert 1999). More specifically, a set of distinctly different networks may all be mapped on to any single given dynamic behavior pattern and, indeed, vice-versa, The same static network may display a spectrum of behaviors (Harush and Barzel 2017). We focus on dynamics rather than network topology.

Abducting a canonical first-order approximation from nonequilibrium thermodynamics (de Groot and Mazur 1984), we can now define a 'real' entropy—as opposed to a 'Shannon uncertainty' that is basically Feynman's (2000) free energy—from the *iterated* free energy F by taking the standard Legendre Transform (Landau and Lifshitz 2007), so that

$$S \equiv -F + ZdF/dZ \tag{8.9}$$

The standard first-order Onsager approximation from nonequilibrium thermodynamics is then (de Groot and Mazur 1984)

$$\partial Z/\partial t \approx dS/dZ = ZdF^2/dZ^2 \tag{8.10}$$

where the scalar diffusion coefficient has been set equal to one.

For a second-order model

$$\begin{aligned} S_2 &= -F + ZdF/dZ + Z^2 d^2 F/dZ^2 \\ \partial Z/\partial t &= dS_2/dZ + ZdS_2^2/dZ^2 \\ &= 3Zd^2F/dZ^2 + Z^2 d^3 F/dZ^3 \\ &\quad + Z\left(3d^2F/dZ^2 + 5Zd^3F/dZ^3 + Z^2 d^4 F/dZ^4\right) \end{aligned} \tag{8.11}$$

8.5 Two Probability Distributions

We consider rates of cognition for two different underlying probability models, according to their underlying 'hazard rates' $Q(x)$ defined as

$$Q(x) = \frac{\rho(x)}{1 - \int_0^x \rho(u)du} \tag{8.12}$$

We choose hazard rates $Q(x) = 1,\ 1/(1 + x)$. The resulting integral equations lead to probability distributions $\rho(x) = \exp[-x],\ 1/(1 + x)^2$. The first is simply the Boltzmann Distribution and the second is 'fat-tailed', without finite mean or variance.

For the Boltzmann Distribution the 'temperature' and cognition rate relations from Eqs. (8.7) and (8.8) are

$$\begin{aligned} g &= \frac{-F}{W(n,\ -F)} \\ L &= \exp[-R_0/g] \end{aligned} \tag{8.13}$$

where $W(n,\ x)$ is the Lambert W-function of order n that satisfies the relation

$$W(n,\ x)\exp[W(n,\ x)] = x$$

It is real-valued only for $n = 0,\ -1$ and $x > -\exp[-1]$. This condition ensures the existence of Fisher Zero phase transitions in the system (Dolan et al. 2001; Fisher 1965; Ruelle 1964 Sect. 5).

For the $1/(1 + x)^2$ distribution

$$\begin{aligned} g &= -F + \frac{1}{2} + \frac{\sqrt{1 - 4F}}{2} \\ L &= \frac{g}{g + R_0} \end{aligned} \tag{8.14}$$

Here, there is also a necessary condition for a real-valued temperature analog, leading again to the possibility of Fisher Zero phase transition.

8.6 Weber-Fechner Implies Yerkes-Dodson

In addition to noise inherent to the RDT approach, suppose there is a 'noise' representing the difficulty of the problem addressed, what Fig. 8.1 identifies as 'impairment of divided attention, working memory, decision-making and multitasking'. That 'noise' is incorporated via the second term in the stochastic differential equations

$$\begin{aligned} dZ_t &= (dS/dZ)dt + \sigma Z_t dB_t \\ dZ_t &= \left(dS_2/dZ + Zd^2S_2dZ^2\right)dt + \sigma Z_t dB_t \end{aligned} \tag{8.15}$$

where σ is the magnitude of the difficulty noise and dB_t is ordinary Brownian noise.

We next replace the $(dZ/dt)dt$ terms with their full expressions in their respective free energy constructs F, to first and second order.

It is here that the Weber-Fechner Law forecloses possibilities. For Eq. (8.1) we set $k = S_0 = 1,\ S = Z$ and assume that perception must be stabilized under the Ito Chain Rule (Protter 2006; Cyganowski et al. 2002). That is, we require that $< d\ln(Z_t) >= 0$ for both first and second order models.

Surprisingly, F, the first, and F_2, the second order, expressions are 'easily' found, using the MAPLE computer algebra system applet of Cyganowski et al. (2002) as

$$\begin{aligned} F(Z) &= \frac{1}{4}\sigma^2 Z^2 + C_1 Z + C_2 \\ F_2 &= C_4 + \frac{1}{24}\sigma^2 Z^2 + C_1 Z + \frac{C_3}{Z} + C_2 \ln(Z) \end{aligned} \tag{8.16}$$

Both expressions have potentially dominant terms in $\sigma^2 Z^2$ that ensure the punctuated emergence of imaginary-valued cognition rates via the algebraic forms of the temperature analogs $g(Z)$.

In what follows, for F, $C_1 = -1,\ C_2 = 3$, and for F_2, $C_4 = 3,\ C_1 = -1,\ C_2 = C_3 = 1$. In the expressions for cognition rate L, we set $n = 0,\ R_0 = 1$ and calculate cognition rates vs. $\{\sigma,\ Z\}$ for both distributions and both orders of approximation. The results are shown in Fig. 8.3. The top two are, respectively, from left to right, first and second order for the Boltzmann Distribution. The bottom two, again left to right, first and second order, for the $1/(1+x)^2$ distribution.

Although there are detailed differences between orders and distribution form results, the Boltzmann with finite mean and variance, and the other fat-tailed and without, the general pattern is that increasing 'noise' σ transforms cognition rate from the simple task of Fig. 8.1 into the inverse-U of the Yerkes-Dodson relation. Note that sufficient 'noise' σ fully collapses all systems. A subsequent section explores this dynamic in deeper detail, focusing on the character of the assumed underlying probability distribution.

In more detail, Fig. 8.4 takes two-dimensional crossections across Fig. 8.3a, setting $\sigma = 0,\ 0.5$, letting Z vary as 'arousal,' and dividing cognition rate into real and imaginary-valued components. At low arousal, both systems suffer a nonzero imaginary-valued 'hallucination' mode. For $\sigma = 0.5$, the intermediate zone, with a zero imaginary valued component, follows the classic inverse-U of the Yerkes-Dodson effect, and imaginary-valued 'panic' emerges at high Z. The $\sigma = 0$ system corresponds to the 'easy' problem of the Yerkes-Dodson effect.

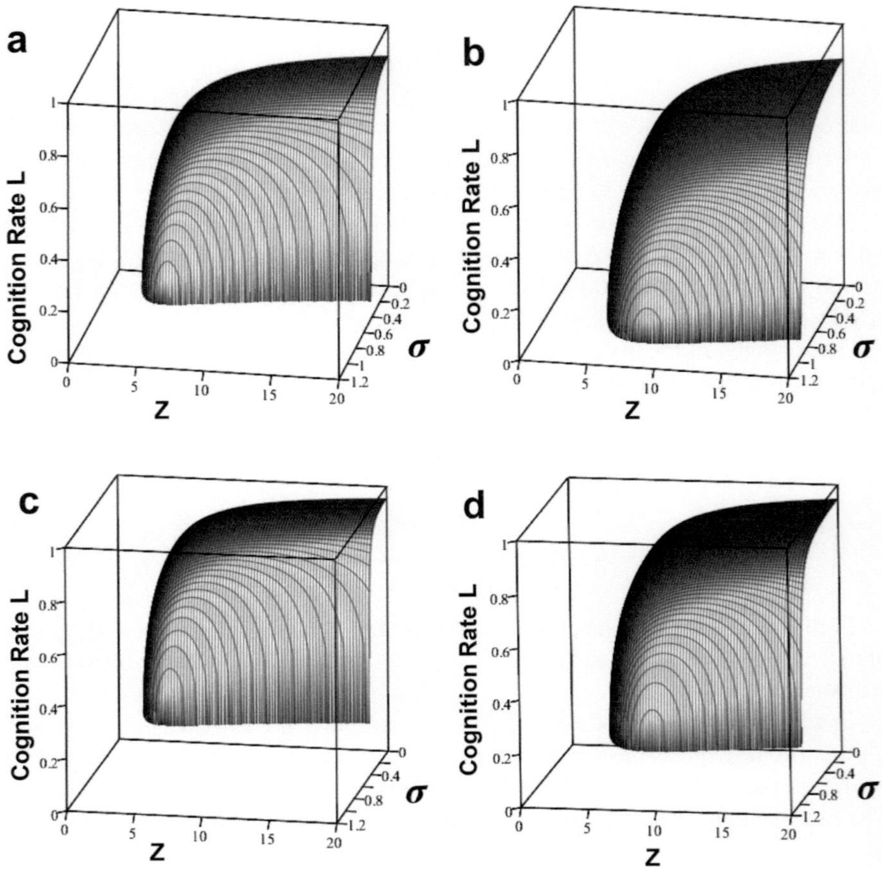

Fig. 8.3 Weber-Fechner compression. Noise-driven transitions of cognition rate from 'easy' to 'difficult' task inverse-U for two orders of approximation and two distributions for the WF law. **a** First order Boltzmann. **b** Second order Boltzmann. **c** First order $1/(1+x)^2$. **d** Second-order $1/(1+x)^2$. Sufficiently large σ collapses all systems

8.7 Stevens Implies Yerkes-Dodson

If, instead of WF data compression according to $< d\ln(Z_t) >= 0$, we impose a version of Stevens Law compression (Dehaene 2003; Krueger 1989; MacKay 1963) as $< dZ_t^{1/n} >= 0,\ n > 1$, then 'elementary' calculation based on the Ito Chain Rule finds

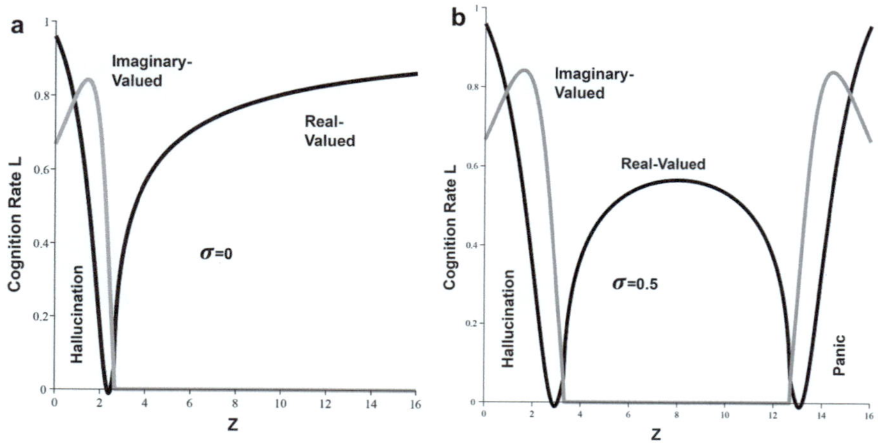

Fig. 8.4 Two-dimensional crossections across Fig. 8.3a, fixing $\sigma = 0,\ 0.5$, with Z as 'arousal.' Cognition rate has been divided into real and imaginary-valued components. At low arousal, both systems show a nonzero imaginary-valued 'hallucination' mode. For $\sigma = 0.5$, 'panic' emerges at sufficiently high arousal, and the intermediate zone, with a zero imaginary component, follows the classic inverse-U. The $\sigma = 0$ system corresponds to the 'easy' problem of the Yerkes-Dodson effect

$$
\begin{aligned}
F(Z) &= \frac{\left(\sigma^2 n - \sigma^2\right) Z^2}{4n} + C_1 Z + C_2 \\
F_2(Z) &= -\frac{C_1}{2Z} - C_2 \ln(Z) + \frac{Z^2\sigma^2}{24} - \frac{Z^2\sigma^2}{24n} + C_3 Z + C_4
\end{aligned}
\tag{8.17}
$$

For large enough n, the results are similar to those following from Eq. (8.16), producing, in Fig. 8.5, for $n = 3$, close analogs to Figs. 8.3 and 8.4, given appropriate boundary conditions. In particular, the condition for the term in Z is again taken as -1, and the constant boundary condition as 3, with the other two again set equal to one.

8.8 Hick-Hyman Implies Yerkes-Dodson

Recall that the Hick-Hyman Law (Sayood 2028) asserts that individual-level response time to a multimodal task challenge increases as the Shannon Uncertainty across the modes of that challenge.

For what we do here, defining $Z(\mathscr{C}, \mathscr{H}, \mathscr{M})$ in terms of appropriate information and 'materiel' streams, the *rate of response* will be determined as $\propto Z^{-1}$. In the formalism of this study, we must now examine the nonequilibrium steady state condition $< dZ_t^{-1} >= 0$, leading to

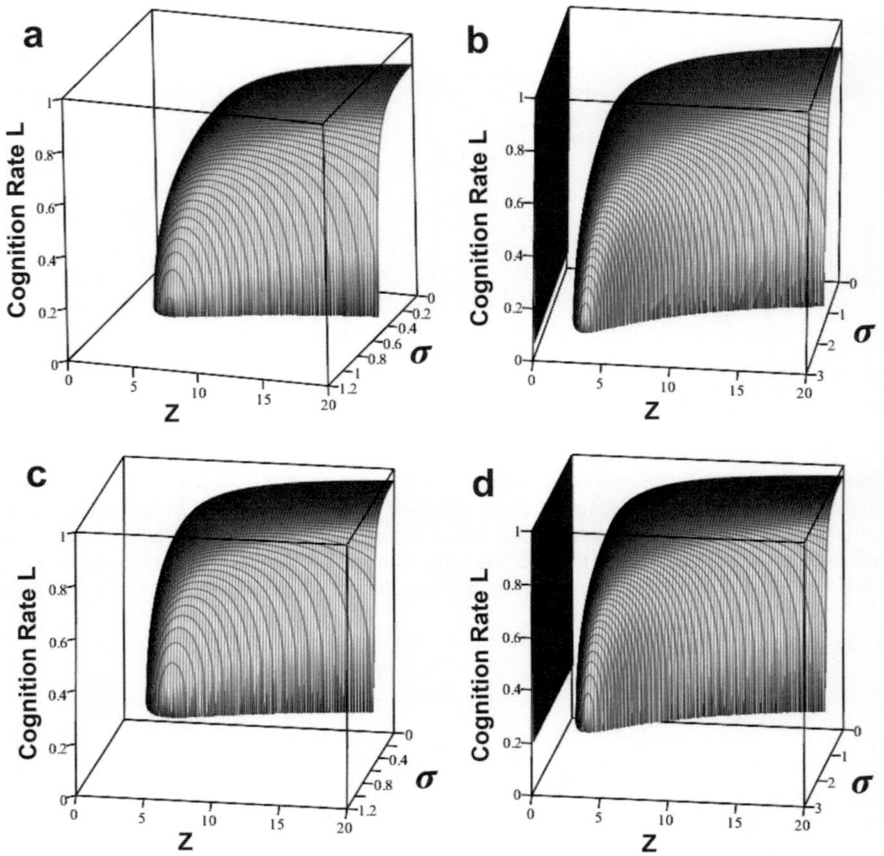

Fig. 8.5 Stevens compression for $n = 3$. Noise-driven transitions of cognition rate from 'easy' to 'difficult' task inverse-U for two orders of approximation and two distributions. **a** First order Boltzmann. **b** Second order Boltzmann. **c** First order $1/(1+x)^2$. **d** Second-order $1/(1+x)^2$. Again, sufficiently large σ collapses the system

$$
\begin{aligned}
F(Z) &= \frac{1}{2}\sigma^2 Z^2 + C_1 Z + C_2 \\
F_2(Z) &= \frac{\sigma^2 Z^2}{12} - \frac{C_1}{2Z} - C_2 \ln(Z) + C_3 Z + C_4
\end{aligned} \tag{8.18}
$$

These relations again produce analogs to Figs. 8.3 and 8.5 in first and second order across the Boltzmann and $1/(1+x)^2$ distributions.

Extension to the relation $< dZ_t^{-k} >= 0,\ k > 0$ is straightforward, producing no surprises, which is, perhaps, in itself surprising.

8.9 Pieron Implies Yerkes-Dodson

The Pieron Law states that system response time will be proportional to the intensity I of the sensory input signal as $T_R \propto I^{-b}, \ b > 0$. Calculation finds, for first and second-order Onsager approximations, letting $I = Z_t$,

$$
\begin{aligned}
F(Z) &= \frac{\sigma^2 (b+1) Z^2}{4} + C_1 Z + C_2 \\
F_2(Z) &= -\frac{C_1}{2Z} - C_2 \ln(Z) + \frac{Z^2 \sigma^2 b}{24} + \frac{\sigma^2 Z^2}{24} + C_3 Z + C_4
\end{aligned}
\tag{8.19}
$$

Given the same boundary conditions as above, one again obtains, for Boltzmann and $1/(1+x)^2$ distributions, close analogs to Figs. 8.3 and 8.4.

8.10 Other Compression Schemes

Suppose we are constructing a cognitive embodied machine, institution, or composite entity, and are able to choose a particular data compression scheme $V(Z)$ under some given environmental or other selection pressure, according to the first and second order 'ordinary volatility' schemes of Eq. (8.15). Then, in first and second orders, the Ito Chain Rule calculation gives

$$
\begin{aligned}
F(Z) &= \int \left(\int -\frac{\sigma^2 Z \left(\frac{d^2}{dZ^2} V(Z)\right)}{2\left(\frac{d}{dZ} V(Z)\right)} dZ \right) dZ + C_1 Z + C_2 \\
F_2(Z) &= \int \left(\int \frac{-\frac{\left(\int \frac{\sigma^2 \left(\int \frac{Z^3 \left(\frac{d^2}{dZ^2} V(Z)\right)}{\frac{d}{dZ} V(Z)} dZ \right) - 2C_1}{Z^2} dZ \right)}{2} + C_2}{Z^2} dZ \right) dZ \\
&\quad + C_3 Z + C_4
\end{aligned}
\tag{8.20}
$$

For appropriate boundary conditions, in all cases, $\sigma = 0$ collapses system dynamics to the 'easy' Y-D problem.

Details for $V(Z_t) = \arctan(Z_t)$ are left as an exercise.

A mathematically sophisticated reader might derive general results across sufficiently draconian compression schemes using methods similar to those of Appleby et al. (2008). Indeed, Eqs. (8.16)–(8.20) can probably be used to bracket most reasonable approaches.

French et al. (1999) were right: data compression provides a possible route to the Yerkes-Dodson Effect.

8.11 The Ductile-Brittle Transition

Figures 8.3, 8.4 and 8.5, as constructed, display disconcerting patterns of instability at low and, under significant noise σ, at high levels of arousal Z. These patterns are, in a sense, 'artifacts' of the boundary conditions imposed in the definitions of the free energy constructs F. Such 'artifacts' haunted materials science and engineering until the latter part of the 20th Century when a broad range of substances, from glasses to ceramics and metals, were understood to undergo sharp temperature-dependent phase transitions from ductile to brittle under sudden stress.

Many readers will be familiar with the standard laboratory demonstration of freezing a normally ductile substance in liquid nitrogen, and then easily shattering it with a light blow. Cognitive systems subject to the Yerkes-Dodson effect, it seems, are likewise subject to kinds of brittleness that can, perhaps, be more formally characterized.

Recall the first expression of Eq. (8.16) for F under the Weber-Fechner Law, and Fig. 8.4a and b, where $C_1 = -1,\ C_2 = 3$ for $\sigma = 0,\ 0.5$, the 'easy' and 'difficult' modes of the Yerkes-Dodson effect. Figure 8.6 displays the analog to Fig. 8.4 for which, again, $C_1 = -1$ but now $C_2 = \exp[-1]$. Hallucination has disappeared.

Boundary conditions, it seems, together define another 'temperature', characterizing the onset of both 'hallucination' and 'panic' modes. Indeed, for Fig. 8.4b, panic begins at $Z = 12.678$, while in Fig. 8.6b, absent hallucination, panic onset is increased to $Z = 16$.

The keys to the matter are found in Eq. (8.16), expressing the form of F under the WF law, and the expressions for g and L in Eq. (8.13) under the Boltzmann distribution. Both g and L depend on the Lambert W-function of orders 0 and -1. Recall that the natures of both g and L are determined by $W(0, -F)$, real-valued only for $-F > -\exp[-1]$. Some elementary algebra, based on the first form of Eq. (8.16), finds the Z-values determining the limits and onset of hallucination and panic given as

$$
\begin{aligned}
Z_H &= -\frac{2\left(C_1 + \sqrt{\sigma^2 e^{-1} - \sigma^2 C_2 + C_1^2}\right)}{\sigma^2} \\
Z_P &= \frac{-2C_1 + 2\sqrt{\sigma^2 e^{-1} - \sigma^2 C_2 + C_1^2}}{\sigma^2}
\end{aligned}
\tag{8.21}
$$

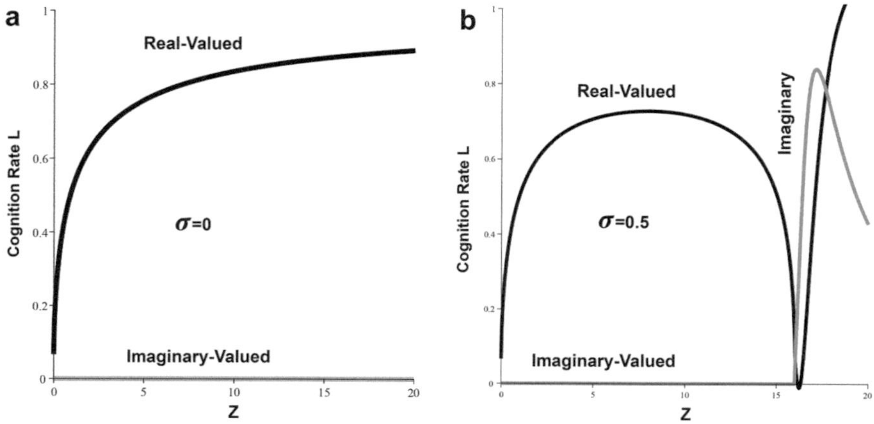

Fig. 8.6 The analog to Fig. 8.4 for which, again, from the first of Eq. (8.16), $C_1 = -1$ but now $C_2 = \exp[-1]$. The boundary conditions act together defining another 'temperature', characterizing the onset of both 'hallucination' and 'panic' modes. For Fig. 8.4b, panic begins at $Z = 12.678$, while here, absent hallucination, panic onset is increased to $Z = 16$ in Fig. 8.6b

Further, solving the relation $Z_P - Z_H = 0$ for σ^2 gives the maximum tolerable level of 'noise' as

$$\sigma^2 = -\frac{C_1^2}{e^{-1} - C_2} \tag{8.22}$$

These relations can 'easily' be extended to various orders of F and to different underlying probability distributions, for example the $1/(1+x)^2$ distribution leading to the expressions of Eq. (8.14).

Bastounis et al. (2021) appear to have been on to something. It seems possible, at least in theory, to limit 'hallucination' in real-time, real-world, embodied cognitive systems. Avoiding panic, however, is quite another matter.

8.12 The Compression/Sensitivity Explosion

We have, thus far, explored systems for which the underlying fundamental distribution was characterized by either a fixed or declining hazard rate function—Eq. (8.12)—under Weber-Fechner or Stevens data compression schemes, to first and second orders. It is conceivable, perhaps even to be expected, however, that some systems may need—or be driven—to increase the rate of signal detection with rising system burden, so that, from Eq. (8.12), $Q(x) = x^m$, $m > 0$. This diktat generates an extended version of the Rayleigh Distribution as

$$\begin{aligned}\rho(m,\, x) &= x^m \exp[-\frac{x^{m+1}}{m+1}] \\ < x > &= (1+m)^{\frac{1}{1+m}}\, \Gamma\left(\frac{2+m}{1+m}\right)\end{aligned} \tag{8.23}$$

Given a free energy measure F, calculation finds the temperature analog and cognition rate as

$$\begin{aligned}g(m,\, F) &= -\frac{F\,(1+m)^{-\frac{1}{1+m}}\left(-\frac{1}{W\left(-\frac{F}{1+m}\right)}\right)^{-\frac{m}{1+m}}}{W\left(-\frac{F}{1+m}\right)} \\ L(m,\, g) &= \mathrm{e}^{-\frac{g^{-1-m}}{1+m}}\end{aligned} \tag{8.24}$$

where, again, W is the Lambert W-function and we have again set the ‘trigger level’ as $H_0 = 1$.

Recall, by contrast, the expressions for the Boltzmann Distribution:

$$g = -F/W(-F),\;\; L = \exp[-1/g].$$

We define F by again imposing Weber-Fechner compression, i.e., stability as $< d\ln(x) >= 0$, according the first-order relation of Eq. (8.16). Figure 8.7a shows, for comparison, the Boltzmann Distribution form of $g(\sigma,\, Z)$, subsequently filtered through the relation $L = \exp[-1/g]$ to produce the cognition rate model of Fig. 8.3a. Figure 8.7b shows $g(m,\, \sigma,\, Z)$ from Eq. (8.24), taking $m = 1/2 \rightarrow 3$ by increments of $1/2$. These results are then fed through the second expression of Eq. (8.24) to give highly complex and counterintuitive variants of Figs. 8.3 and 8.4 that are left as an exercise.

Requiring, or imposing, rising signal detection rates with increasing signal strength, under conditions of signal compression and ‘noise’, can produce incomprehensively complicated cognitive dynamics.

Somewhat heuristically, in signal detection systems, sensitivity is often represented by a threshold value chosen above the mean noise level. As sensitivity increases, the system becomes more responsive to smaller changes in signal strength. This can be visualized as lowering the threshold for detection, allowing weaker signals to be recognized. Combining increasing sensitivity with data compression schemes like the Weber-Fechner law creates a system that is highly responsive to small changes in input while also compressing the range of those inputs. This combination can lead to instabilities for several reasons:

- Amplification of Noise: As sensitivity increases, the system becomes more susceptible to detecting noise as signal, potentially leading to false positives.
- Compression of Dynamic Range: The Weber-Fechner law—like similar schemes—compresses the perceived intensity of stimuli, which can make it

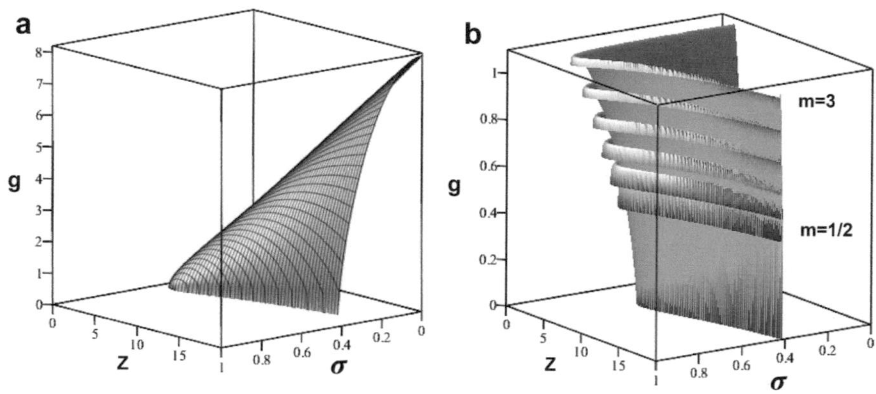

Fig. 8.7 **a** $g(\sigma, Z)$ for the Boltzmann distribution, producing the relatively simple Y-D cognition rate model of Fig. 8.3a. **b** The hazard rate model $Q(x) = x^m$, showing $g(m, \sigma, Z)$ from Eq. (8.24). Here, $m = 1/2 \rightarrow 3$ by increments of $1/2$. These versions of g are then fed through the second expression of Eq. (8.24) to give highly irregular variants of Fig. 8.3

difficult for the system to distinguish between important signals and background noise at higher intensities.

- Feedback Loops: In complex systems with multiple interconnected modules, increased sensitivity can create feedback loops that amplify small fluctuations, potentially leading to system-wide instabilities that go far beyond those illustrated by Fig. 8.7

8.13 Extending the Models

Volatility

So far we have, unfortunately, only chipped away at the tip of a very large formal iceberg. If we—in first order—extend the perspective on 'volatility', the basic equations of interest become

$$\begin{aligned} dZ_t &= \left(Z_t d^2F/dZ^2\right) dt + \sigma g(Z_t) dB_t \\ < dV(Z_t) > &= 0 \end{aligned} \tag{8.25}$$

where, again, $F(Z)$ is the first-order free energy analog. Here, however, $g(Z)$ represents a generalized volatility function and $V(Z)$ characterizes the mode of 'sensory' signal compression.

The general form of F, via the Ito Chain Rule, is then 'easily' found to be

$$F(Z) = \int \left(\int -\frac{\sigma^2 g(Z)^2 \left(\frac{d^2}{dZ^2} V(Z)\right)}{2Z \left(\frac{d}{dZ} V(Z)\right)} dZ \right) dZ + C_1 Z + C_2 \tag{8.26}$$

so that $\sigma = 0$ produces the 'easy' form of the Yerkes-Dodson effect for appropriate boundary conditions.

As an example we replace the 'simple' volatility term in the first of Eq. (8.15) with $\sigma\sqrt{Z_t}dB_t$—again taking $V(Z_t) = \ln(Z_t)$—for the Boltzmann and $1/(1+x)^2$ distributions. This leads to the cognition rate relations of Fig. 8.8a and b, which should be compared to Fig. 8.3a and c.

Evidently, the exact form of 'volatility' is of considerable importance.

The Tyranny of Time

A second direction for extension of the formalism involves the tyranny of time. That is, not only are the individual components of the composite rate index Z constrained, but the time available for effective action is limited.

We address this in a standard—and admittedly crude—first-order approximation as a classic Lagrangian optimization under environmental shadow price constraints abducted from economic theory (Jin et al. 2008; Robinson 1993). Here, time and (other) resource limits strongly dominate system dynamics.

We begin with an attempt to maximize cognition rate across a full system of n subcomponents. Each has an associated cognition rate L_j and is alloted resources at the rate Z_j for time T_j. We construct a Lagrangian as

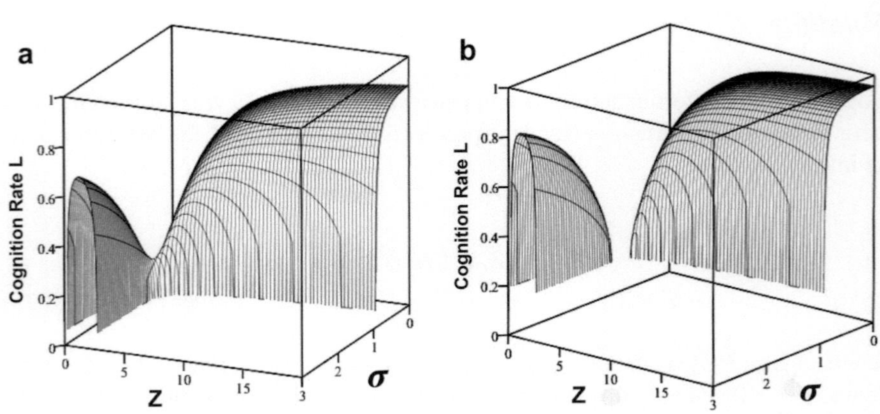

Fig. 8.8 Alterations in the cognition rate examples of Fig. 8.3a and c consequent on replacement of the 'volatility' $\sigma Z_t dB_t$ with $\sigma\sqrt{Z_t}dB_t$

$$
\begin{aligned}
\mathscr{L} &\equiv \sum_i L_i + \lambda\left(Z - \sum_i Z_i\right) + \mu\left(T + \sum_i T_i\right) \\
\partial\mathscr{L}/\partial Z &= \lambda, \ \partial\mathscr{L}/\partial T = \mu \\
\partial L_i/\partial Z_i &= \lambda, \ \partial L_i/\partial T_i = \mu
\end{aligned} \tag{8.27}
$$

Again, from economic theory, λ and μ are shadow prices imposed by environmental externalities including fog, friction, and usually skilled adversarial intent.

Some heuristic manipulation gives

$$
\partial Z_i/\partial T_i = \frac{\mu}{\lambda} \equiv \omega \tag{8.28}
$$

Taking $g(Z_i)$ as the volatility function of the previous subsection, we can write the fundamental stochastic differential equation under draconian constraints of time and resource rates as

$$
dZ_t^i = \omega dt + \sigma_i g_i(Z_t^i) dB_t \tag{8.29}
$$

If the data compression function is $V(Z) = \ln(Z),\ Z^{-1},\ Z^{1/n}$, and so on, then application of the Ito Chain Rule to determine the nonquilibrium steady state condition $< dV(Z_t^i) >= 0$ gives, suppressing indices,

$$
\begin{aligned}
\frac{2\omega}{\sigma^2} &= -\frac{g(Z)^2\left(\frac{d^2}{dZ^2}V(Z)\right)}{\frac{d}{dZ}V(Z)} \equiv Q(Z) \\
Z_i &= Q_i^{-1}\left(2\omega/\sigma^2\right)
\end{aligned} \tag{8.30}
$$

Typically, $Q^{-1} \to 0$ as σ increases or ω declines.

For the three forms of $V(Z)$ given just above, ‘it is easy to show’ that, under ordinary volatility, taking $g_i(Z_i) = Z_i$,

$$
Z_i = \frac{2\omega}{\sigma^2},\ \frac{\omega}{\sigma^2},\ \frac{2\omega n}{\sigma^2(n-1)} \tag{8.31}
$$

Declining environmental shadow price ratio is synergistic with rising ‘noise’ to drive essential resource rates below critical values for important subcomponents.

Ineeed, if we, by fiat, impose ordinary volatility, then a general result emerges across all possible compression schemes $V(Z)$.

Rewriting Eq. (8.30),

$$\begin{aligned}\frac{2\omega}{\sigma^2} &\equiv \kappa = -\frac{Z^2 d^2 V(Z)/dZ^2}{dV(Z)/dZ} \\ V(Z) &= C_1 + \left(-\frac{Z\,e^{\frac{\kappa}{Z}}}{\kappa} - \mathrm{Ei}_1\left(-\frac{\kappa}{Z}\right)\right) C_2\end{aligned} \tag{8.32}$$

where Ei_1 is the exponential integral of order 1.

Multiplying both sides of the second relation by κ,

$$\kappa V(Z) = \kappa C_1 - C_2 Z\,e^{\frac{\kappa}{Z}} - \kappa C_2 \mathrm{Ei}_1\left(-\frac{\kappa}{Z}\right) \tag{8.33}$$

Taking the limit $\kappa \to 0$, by inspection, gives $0 = -C_2 Z$, so that, for 'ordinary volatility', under any compression scheme $V(Z)$, essential subcomponents will be driven to extinction by either a sufficiently small shadow price ratio, or by enough 'noise'.

The Tyranny of Evolution

A related extension is described by Wirtz (2024b) as follows:

> Standard operating procedures that continue for years or even decades give opponents the time they need to devise innovative tactics, technologies, and stratagems that are hard to detect, analyze, and forecast. Small changes can delay opponents' schemes, while larger changes can invalidate their plans altogether. Routine can be exploited. It is possible for planners and commanders to minimize the challenge facing intelligence by providing opponents with a new problem to plan against before they devise a way to solve the old one. And rest assured, they are working on finding ways to sidestep current force postures.

That is, while institutional cognition rates are most often perceived by combatatants/participants on tactical and operational levels and relatively short time frames, strategic enterprise takes place over Darwinian and Lamarckian evolutionary spatial, social, and time scales. Wallace (2020, Chaps. 1 and 9) explores the dynamics of such evolutionary process in terms of 'punctuated equilibrium' phase transitions analogous to the Fisher Zero phase changes studied here. However, extension of the cognitive dynamics models studied here to evolutionary time scales is also possible, following Wallace (2023, Sect. 2.5).

Central to gene frequency evolutionary models and the Evolutionary Extended Synthesis extensions is the idea of optimization on a 'fitness landscape' or on paths constrained within such a landscape (e.g., Gonzalez-Forero and Gardner 2021, 2022). More generally, one can seek to maximize some scalar index of critical function across a multi-component system constrained by resource rates and by the necessity of responding collectively to environmental signals or changes within time constraints. A central feature is an assumption that the rate at which an essential resource is delivered to component i is governed—in first order—according to Eq. (8.10) as

$$\partial Z_i/\partial t = dS/dZ_i = f_i(Z_i) \tag{8.34}$$

The full system is seen as consisting of $i = 1, 2, \ldots, m$ cooperating units that must be individually supplied with resources at rates Z_i under the overall constraints of resources and time as $\sum_i Z_i = Z$, $\sum_i T_i = T$. We then optimize some critical scalar index of function across the full system under these constraints.

As an example, we take the sum of subcomponent cognition/reaction rates of Eq. (8.8) and variants, assuming $dZ_i/dt = f(Z_i) \rightarrow 0$.

While more general approaches to optimization than simple Lagrangian are possible (e.g., Nocedal and Wright 2006), here, as above, we use a general Lagrangian optimization across subcomponent cognition/reaction rates as

$$\begin{aligned}
\mathcal{L} &= \sum_i L_i + \lambda\left(Z - \sum_i Z_i\right) + \mu\left(T - \sum_i T_i\right) \\
\partial\mathcal{L}/\partial Z &= \lambda \\
\partial L_i/\partial Z_i &= \lambda \\
\partial L_i/\partial T_i &= \mu
\end{aligned} \tag{8.35}$$

remembering in particular that $\partial Z_i/\partial T_i = f_i(Z_i)$.

Again, for simple economic models, λ and μ represent the 'shadow prices' imposed on system dynamics by environmental constraints, in a large sense (Jin et al. 2008; Robinson 1993).

While many such formalisms are possible, elementary Lagrangian optimization outlines the basic mechanisms.

Algebraic manipulation—which will apply to virtually all possible scalar function measures—gives

$$\begin{aligned}
f_i(Z_i) &= \frac{\mu}{\lambda} \\
Z_i &= f_i^{-1}(\frac{\mu}{\lambda})
\end{aligned} \tag{8.36}$$

so that a monotonic function $f_i(Z_i)$ generates a monotonic dependence of Z_i on the shadow price ratio.

For example, setting $f_i(Z_i) = \beta_i - \alpha_i Z_i$—an 'exponential' model—produces the expression

$$Z_i = \frac{\beta_i - \mu/\lambda}{\alpha_i} \tag{8.37}$$

As above, sufficient environmental shadow price—selection pressure—will drive any component resource rate Z_i below what is needed for critical modular function.

Again, shadow prices characterize environmental burdens—selection pressures—leading to decline in the maximum possible level of resource delivery Z_{max} below some critical value Z_{crit}. Optimization becomes impossible, and we could then write both Z_{max} and Z_{crit} for a critical subsystem as appropriate functions of some integral of the environmental shadow price over the duration of acute selection pressure. Details will vary across 'organisms', institutions, and environments, with optimization model and measures of λ, μ and Z, but the mechanism seems canonical.

Environmental shadow prices impose selection pressure for any possible scalar measure of fitness in this formulation.

Multivariate versions of the model, in which, for example, $Z \to Z^j$, $\lambda \to \lambda^j$, $j = 1, 2, \ldots$ and/or $T \to T^k$, $\mu \to \mu^k$, $k = 1, 2, \ldots$, lead toward the classic 'fitness landscapes' described by Gavrilets (2010) and references therein. The shadow price parameters, and their relations, then characterize landscape topological roughness.

'Rough landscape' optimization bedevils a broad range of engineering, scientific, and practical enterprise, most particularly recent developments in machine learning and artificial intelligence, leading to far realms of topology, applied mathematics, and algorithm design.

The dynamics of system fragmentation under sufficient selection pressure, including estimation of fragment size after a 'speciation event', can be found in Chap. 6 of Wallace (2023).

8.14 Discussion

An art of science is inference of the general from the particular. The models of embodied cognition studied here suggest that necessary compression of sensory/intelligence and related internal data streams can drive the Yerkes-Dodson Effect's contrast between 'easy' and 'difficult' tasks by such things as increasing impairment of divided attention, limits on working memory, difficulties in decision-making, and the burdens of multitasking (again, Diamond et al. 2007; Fricchione 2023). Further, the inverse-U patterns of difficult problems appear routinely bracketed by hallucination at low, and panic at high, arousal, depending critically on the interaction between 'boundary conditions' and the basic underlying probability distribution or distributions.

Another art-of-science, however, is recognizing the severe limitations of mathematical modeling in the study of complex real-world phenomena. As Pielou (1977), argues in the context of theoretical ecology, the principal utility of mathematical models is speculation, the raising of questions to be answered by the analysis of observational and experimental data, the only source of new knowledge.

This being said, similar arguments have been made regarding the dynamics of organized conflict on Clausewitz landscapes of fog, friction, and deadly adversarial intent (e.g., Wallace 2020, 2021a, 2021b, 2024, and the many classic and classical references therein). An underlying mechanism, then, seems related to sufficient—and usually badly needed—compression scaling of sensory/intelligence information data rates, enabling Maskirovka, spoofing, and related deceptions. Internal data streams are, likewise, often compressed, leading to the synergisms explored in Figs. 8.3, 8.4 and 8.5 and Fig. 8.7, although suggesting the possibility of some mitigation of 'hallucination' in Fig. 8.6 and Eqs. (8.21) and (8.22). The argument leading to Fig. 8.8 further suggests the importance of the exact way in which 'noise' influences system dynamics.

Another mechanism inherent to such conflict, however, is paralysis of action by uncertainty, the Hick-Hyman Law, and by 'low signal', the Pieron Law. Both under-supply and over-burden of intelligence/sensory information lead to catastrophe by the Yerkes-Dodson effect, bracketed by hallucination and panic.

Most critically, however, Fisher Zero phase transitions implied by Eqs. (8.13), (8.14), and the optimization models, suggest that inverse-U signal transduction and associated patterns of hallucination, panic, or full system fragmentation and extinction, will not be constrained to these circumstances. We have explored only a few trees in a very large forest.

In addition—and perhaps centrally—new probability models of poorly-understood complex phenomena, if solidly based on appropriate asymptotic limit theorems, can serve as the foundation for building new, robust, statistical tools useful in the analysis, and modest control, of those phenomena.

The development, testing, and validation of such statistical tools, however, is not a project for the timid.

References

Appleby, J., X. Mao, A. Rodkina, 2008, Stabilization and destabilization of nonlinear differential equations by noise, IEEE Transactions on Automatic Control 53:683–691.

Barabasi, A., R. Albert, 1999, Emergence of scaling in random networks, Science 286:509–512.

Bastounis, A., A. Hansen, V. Vlacic, 2021, Why deep learning is unstable despite the existence of stable neural networks, arXiv:2109.06098v1.

Bennett, C.H., 1982, The thermodynamics of computation, International Journal of Theoretical Physics 21:905–940.

Cover, T., J. Thomas, 2006, Elements of Information Theory, Second Edition, Wiley, New York.

Cyganowski, S., P. Kloeden, J. Ombach, 2002, From Elementary Probability to Stochastic Differential Equations with MAPLE, Springer, New York.

de Groot, S., P. Mazur, 1984, Nonequilibrium Thermodynamics, Dover, New York.

Dehaene, S., 2003, The neural basis of the Weber-Fechner law: a logarithmic mental number line, TRENDS in Cognitive Sciences, 7:145–147.

Derman, E., M. Miller, D. Park, 2016, The Volatility Smile, Wiley, New York.

Dolan, B., W. Janke, D. Johnston, M. Stathakopoulos, 2001, Thin Fisher zeros, Journal of Physics A 34:6211–6223.

Feynman, R., 2000, Lectures on computation,Westview Press, New York.
Fisher, M., 1965 Lectures in Theoretical Physics vol. 7, University of Colorado Press, Boulder.
French, V., E. Anderson, G. Putman, T. Alvager, 1999, The Yerkes-Dodson law simulated with an artificial neural network, Cognitive Systems 5:136–147.
Fricchione, G., 2023, Mind body medicine: a modern bio-psycho-social model forty-five years after Engel, BioPsychoSocial Medicine, 17:12.
Gavrilets, S., 2010, High-dimensional fitness landscapes and speciation. In Pigliucci, M., and G. Muller (eds.), Evolution: The Extended Synthesis, MIT Press, Cambridge MA.
Gonzalez-Forero, M., A. Gardner, 2021, A mathematical framework for evo-devo dynamics, bioRxiv preprint https://doi.org/10.1101/2021.05.17.444499.
Gonzalez-Forero, M., A. Gardner, 2022, How development affects evolution, bioRxiv preprint https://doi.org/10.1101/2021.10.20.464947.
Harush, U., B. Barzel, 2017, Dynamic patterns of information flow in complex networks, Nature Communications 8:2181.
Jin, H., Z. Hu, X. Zhou, 2008, A convex stochastic optimization problem arising from portfolio selection, Mathematical Finance 18:171–183.
Kandel, E., T. Jessell, J. Schwartz, S. Siegelbaum, A. Hudspeth, 2013, Principles of Neural Science (5th ed.), McGraw-Hill, New York. p. 451.
Khinchin, A., 1957, Mathematical Foundations of Information Theory, Dover, New York.
Krueger, L., 1989, Reconciling Fechner and Stevens: Toward a unified psychophysical law, Behavioral and Brain Sciences 12:251–267.
Laidler, K., 1987, Chemical Kinetics, 3rd ed, Harper and Row, New York.
Landau, L., E. Lifshitz, 2007, Statistical Physics, 3rd Ed. Part 1, Elsevier, New York.
MacKay, D., 1963, Psychophysics of perceived intensity: A theoretical basis for Fechner's and Stevens' laws, Science, 139:1213–1216.
Nair, G., F. Fagnani, S. Zampieri, R. Evans, 2007, Feedback control under data rate constraints: an overview, Proceedings of the IEEE 95, 108138.
Nocedal, J., S. Wright, 2006, Numerical Optimization, Second Edition, Springer, New York.
Ochiai, Y., 2011, US Intelligence and the Origins of the Vietnam War, 1962-1965, Thesis, University of Edinburgh, https://era.ed.ac.uk/bitstream/handle/1842/5511/Ochiai2011.pdf?isAllowed=y&sequence=2.
Pielou, E.C., 1977, Mathematical Ecology, Wiley, New York.
Protter, P., 2006, Stochastic Integration and Differential Equations, Second Edition, Springer, New York.
Reina, A., T. Bose, V. Trianni, J. Marshall, 2018, Psychophysical laws and the superorganism, Scientific Reports, 8:4387, https://doi.org/10.1038/s41598-018-22616-y.
Robinson, S., 1993, Shadow prices for measures of effectiveness II: General model, Operations Research 41:536–548.
Ruelle, D., 1964, Cluster property of the correlation functions of classical gases, Reviews of Modern Physics, April, 580–584.
Sayood, K., 2028, Information theory and cognition: A review, entropy, 20, 706; https://doi.org/10.3390/e20090706.
Spiegelhalter, D., 2024, Does probability exist? Nature 636:560–563.
Taleb, N. N. (ed.), 2020, Statistical Consequences of Fat Tails: Real world preasymptotics, epistemology, and applications, STEM Academic Press.
Wallace, R., 2020, Cognitive Dynamics on Clausewitz Landscapes: The control and directed evolution of organized conflict, Springer, New York.
Wallace, R., 2021a, How AI founders on adversarial landscapes of fog and friction, Journal of Defense Modeling and Simulation, https://doi.org/10.1177/1548512920962227.
Wallace, R., 2021b, Carl von Clausewitz, the Fog-of-War, and the AI Revolution: The real world is not a game of Go, Springer, New York.
Wallace, R., 2023, Essays on the Extended Evolutionary Synthesis: Formalizations and Expansions, Springer, New York.

Wallace, R., 2024, Fog, friction, and failure in organized conflict: A formal study, Axioms, https://doi.org/10.3390/axioms13020111.
Watts, D., S. Strogatz, 1998, Collective dynamics of small world networks, Nature 393:440–442.
Wikipedia 2025a https://en.wikipedia.org/wiki/Weber%E2%80%93Fechner_law.
Wikipedia 2025b https://en.wikipedia.org/wiki/Stevens%27s_power_law.
Wirtz, J., 2024a, Are intelligence failures still inevitable? International Journal of Intelligence and Counterintelligence 37:307–330, https://doi.org/10.1080/08850607.2023.2214328.
Wirtz, J., 2024b, The Battle of Midway: Five intelligence takeaways for today, Proceedings of the US Naval Institute 150/12/1462 https://www.usni.org/magazines/proceedings/2024/december/battle-midway-five-intelligence-takeaways-today.
Woocher, L., 2011, Conflict assessment and intelligence analysis: Commonality, convergence, and complementarity, https://ciaotest.cc.columbia.edu/wps/usip/0022842/f_0022842_18725.pdf.

Chapter 9
Mission Command

Abstract Those accustomed to acting within 'normal' bureaucracies will have experienced the degradation, distortion, and stunting imposed by inordinate levels of hierarchical 'decision structure,' particularly under the critical time constraints so fondly exploited by John Boyd and his followers. Here, via an approach based on the asymptotic limit theorems of information and control theories, we explore this dynamic in detail, abducting ideas from the theory of carcinogenesis. The resulting probability models can, with some effort, be converted into new statistical tools for analysis of real time, real world data involving cognitive phenomena and their dysfunctions across a considerable range of scales and levels of organization.

A favorable situation will never be exploited if commanders wait for orders. The highest commander and the youngest soldier must always be conscious of the fact that omission and inactivity are worse than resorting to the wrong expedient.

—Von Moltke the Elder, as quoted in Dupuy (1977).

Gentlemen, I demand that your divisions completely cross the German borders, completely cross the Belgian borders and completely cross the river Meuse. I don't care how you do it, that's completely up to you.

—Kurt Zeitzler, Chief of Staff, Panzergruppe Kleist, 13 May 1940.

The Wehrmacht's ability to generate ad hoc battlegroups from fragments of units, to counterattack strongly even before the precise nature of a threat had been identified, and the tempo its units often demonstrated all indicate a loose, flexible and decentralized command philosophy.

—Storr (2007).

9.1 Introduction

In recent Western intellectual tradition, 'Management Strategy' has been the focus of two contrasting, but perhaps complementary, schools of thought, characterized by the works of Mintzberg (1973, 1989) and Porter (1980, 2008).

R. Wallace, *Hallucination and Panic in Autonomous Systems*, Studies in Computational Intelligence 1217, https://doi.org/10.1007/978-3-031-95207-4_9

Mintzberg centers on the 'roles' and 'activities' of managers in their day-to-day work. He identified ten managerial roles grouped into three categories: interpersonal (e.g., leader, liaison), informational (e.g., monitor, disseminator), and decisional (e.g., entrepreneur, negotiator). His work emphasizes the multifaceted and dynamic nature of management, highlighting that managers must balance multiple roles simultaneously to adapt to organizational needs.

His framework is practical for managers seeking to improve their effectiveness by understanding and balancing their roles. It is particularly useful for leadership development and organizational behavior studies.

Mintzberg challenges traditional views of management as a linear process of planning, organizing, leading, and controlling. Instead, he portrays management as a dynamic interplay of roles that require adaptability to complex situations.

Porter focuses on 'strategic management' and achieving competitive advantage. His frameworks, such as the Five Forces Model and value chain analysis, help organizations analyze industry dynamics and develop strategies to outperform competitors. Porter emphasizes long-term strategic positioning through cost leadership, differentiation, or focus strategies.

Porter's models are primarily used by organizations to analyze industries, evaluate competition, and formulate strategies for market positioning. They are widely applied in business strategy and competitive analysis.

Porter adopts an economic perspective, focusing on how firms can create value by leveraging competitive forces. His work is rooted in the idea that strategic choices determine a firm's success in its environment.

Military enterprise constitutes a form of organized conflict that can differ materially from current neoliberal visions of Schumpeterian market dynamics. Indeed, the long history of various imperial defeats in Afghanistan, and the egregious failures of the USA in the Korean and and Vietnam Wars, suggest that God may not, indeed, always be on the side of the bigger—or more technologically advanced—batallions. The term 'doctrine' intrudes as a kind of Lamarckian 'genetic' determinant of tactical and operational dynamics. Strategic matters are left to higher, usually grossly incompetent, authority often both driven and constrained by narrow ideology. Russia's debacle in the Ukraine, as the Appendix explores, provides a contemporary case history. The experience of the Wehrmacht in the USSR supplies historical perspective.

This being said, Mintzberg's emphasis on adaptability to complex situations and Porter's focus on strategic choice broadly resonate with what we will do here, aimed at the development of new probability models of an embedding institutional cognition and its dysfunctions that might be converted into robust and validated statistical tools for the study, and in some measure, the control and remediation, of institutional process.

However, in some contrast, institutional decision-making—essentially institutional cognition in the sense of Atlan and Cohen (1998)—as presented here, unlike Mintzberg's and Porter's work, does not elevate 'management' as a particular salient object central to understanding institutional dynamics. Rather, such a perspective places 'managers' soundly within the confines of doctrine, culturally-confined path-dependent historical trajectory, and the constraint that 'we know this but we can't say this'.

In short, we attempt to move beyond the classic Western cognitive pattern of over-focus on 'salient points' at the expense of context and contextual dynamics (Nisbett et al. 2001). The role of 'managers' is, we assert, precisely one of those overfocused points. Corporate culture, as a whole, remains an all-powerful determinant of institutional—as opposed to individual—cognitive dynamics, in partnership with the 'enemy that has a vote' as Sun Tzu remarks. Indeed, a cardinal rule for advancement in an institutional context is the admonition 'to get along, go along'. What individual 'managers' actually think often has little effect on what they ultimately do. Humans are, after all, the naked mole rats of hominid species. What comes to mind is the line officers ordering soldiers to advance against barbed-wire protected machine gun emplacements at the battle of the Somme.

As Louis Gerstner Jr., former CEO of IBM put it, (Gerstner 2003) "Culture is everything", and he should know.

Although the 'Auftragstaktik' that defeated France so handily in 1940 did not—ultimately—fare well a year later in the Wehrmacht's ill-advised reprise of Napoleonic grandiosity in Russia, circumstances in which such 'mission command' provides decisive advantage continue to abound at the tactical and operational scales of organized conflict (e.g., Storr 2007). More recently, Matzenbacher (2018) characterizes the approach according to the schematic of Fig. 9.1.

Here, we provide a formal analysis of this and other, more convoluted, decision processes, using newly-developed models of cognitive dynamics based on the asymptotic limit theorems of information and control theories, exploring a surprising parallel between hierarchical decision-making and the staged nature of cancer progression. Those whose careers have been deeply embedded in conventional hierarchical organizations are likely to experience a visceral recognition of the analogy.

We begin with some considerable methodological boilerplate.

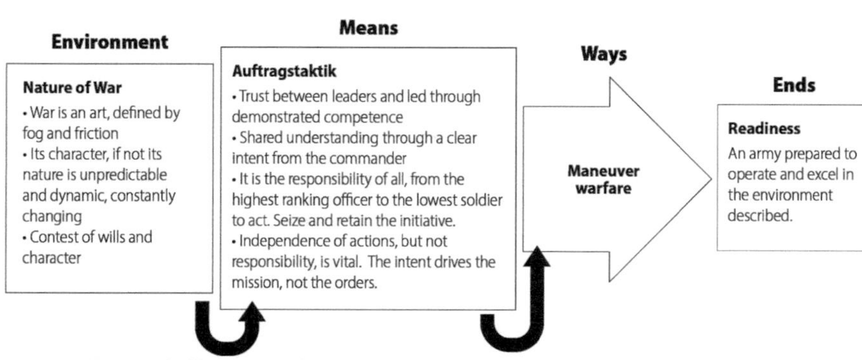

Fig. 9.1 Adapted from Matzenbacher (2018). German army *Auftragstaktik* logic map

The Asymptotic Limit Theorems

A surprisingly limited set of asymptotic limit theorems from information and control theories constrain all possible cognitive structures and their dynamics (Khinchin 1957; Cover and Thomas 2006; Nair et al. 2007):

- The Shannon Coding Theorem that prescribes how to code a message in 'typical' sequences for maximum error-free transmission through a particular channel at a maximum 'channel capacity' (and the 'tuning theorem' variant in which the message formally 'transmits' the channel).
- The Source Coding Theorem that splits a set of sequences emitted by an information source into two subsets, a relatively small one that is 'meaningful'—the sequences are consonant with an underlying 'grammar' and 'syntax'—and a much larger 'meaningless' one 'of measure zero'. Only for stationary, ergodic systems is it possible to characterize 'information' and 'information source uncertainty' in terms of a Shannon entropy-analog. For nonergodic systems, each (infinite) signal sequence has it's own scalar source uncertainty that cannot be expressed in 'Shannon entropy' form (Khinchin 1957). It is here that methods must be abducted from statistical physics, introducing as well equivalence-classes of sequences forming a groupoid, leading to groupoid symmetry-breaking phase transitions characterizing 'cognitive' phase transitions, quite unlike those of 'conventional' physical systems.
- The Rate Distortion Theorem (RDT) and an associated Rate Distortion Control Theory that, in a sense, 'tunes' the channel capacity around a maximum-permissible scalar measure of average distortion (which finesses the Friston 'free energy minimization' approach).
- The Data Rate Theorem (DRT) stating that, for stability of an inherently unstable control system, 'control information' must be provided at a rate greater than the system produces it's own 'topological information'. A version of this is, somewhat surprisingly, 'easily' derived from a time and 'free energy rate' optimization based on the Hartley-Shannon Theorem.

Further development leads to the following:

- Cognition implies choice—of a small set of actions from a larger embedding set of what is possible to the system of interest. Such choice reduces uncertainty in a formal manner, implying the existence of an information source 'dual' to the cognitive process studied, at and across the varied scales and levels of organization of the living state (Atlan and Cohen 1998; Dretske 1994; Maturana and Varela 1980).
- Real-world, real-time cognition is always inherently unstable under the impact of the 'topological information' imposed by embedding environmental 'roadways'. Invocation of the DRT then implies that *cognition must always be paired with regulation for system stability and long-term function under selection pressures*. Further, *it can be argued that the degree of pairing between cognition and regulation is, in fact, a good measure of a system's 'self awareness', a cardinal criterion in many contemporary definitions of consciousness.*

- Here, we further assume an adiabatically, piecewise, stationary (APS) 'Born Oppenheimer' model in which the system of interest remains 'close enough' to stationary during the period(s) of interest so that the asymptotic limit theorems are effectively 'anytime algorithms' (e.g., Zilberstein 1996). Between 'pieces' for which this assumption is sufficiently true, we invoke (groupoid) symmetry-breaking and/or Fisher Zero (Dolan et al. 2001) phase transitions.
- The ergodic restriction is 'easily' lifted, at the cost of imposing methods abducted from statistical physics, with dynamics approximated via an analog to the first (and higher) order Onsager treatment of nonequilibrium thermodynamics, recognizing, however, that, due to the failure of information source microreversibility—e.g., 'eht' does not have the same probability as 'the' in English—there are no 'Onsager reciprocal relations' in high order approximations.
- For purely physical phenomena, free energy is often defined in terms of Boltzmann distribution-derived partition functions. In contrast, free energy analogs associated with the dynamics of cognitive phenomena can, and may well, be characterized by 'fat-tailed' distributions that do not have well-defined first or second moments (e.g., Derman et al. 2016; Wallace and Fricchione 2024).

Resource Rates

Embodiment recognizes that cognitive systems—biological, institutional, mechanical, or composite—are, in reality, ultimately embedded in an environment that includes themselves. There will typically be (at least) three resource streams necessary for their function:

- The first is the rate at which elements of the system can communicate with each other, instantiated by some channel capacity $\mathscr{C}$.
- The second is the rate at which 'sensory information' is available from the embedding circumstance, say according to a channel capacity $\mathscr{H}$.
- Third, there is the rate at which 'material/materiel' resources can be provided, a rate $\mathscr{M}$. For an organism, this might be measured by the rate at which metabolic free energy is provided. Machine/institutional systems will have a different measure.

These three rates will cross-correlate: one is confronting a 3×3 (or larger) matrix $\mathbf{Z}$. An $n \times n$ matrix will have will have n scalar invariants determined by the polynomial relation

$$\begin{aligned} p(\gamma) &= \det[\mathbf{Z} - \gamma \mathbf{I}] \\ &= (-1)^n \gamma^n + (-1)^{n-1} r_1 \gamma^{n-1} + \ldots - r_{n-1}\gamma + r_n \end{aligned} \tag{9.1}$$

$\mathbf{I}$ is the appropriate identity matrix, 'det' is the determinant, and γ is a real-valued parameter. The first invariant is the matrix trace, and the last the determinant. We seek to construct a scalar resource rate measure from these invariants.

The most direct such index might be $Z = \mathscr{C} \times \mathscr{H} \times \mathscr{M}$, but our worlds are not likely so simple.

We postulate—in first-order—that a single scalar index $Z(r_1, \ldots, r_n)$ can be constructed that sufficiently represents information and material resource rates. This approach is analogous to, but different from, the more usual principal component analysis based on a correlation matrix. Wallace (2021) outlines the complications associated with generalization to more than a single such index.

Rate Distortion Control Theory

A directly 'embodied' formalism emerges via a 'Rate Distortion Control Theory' based on the Rate Distortion Theorem. As described, the RDT addresses transmission under conditions of noise over an information channel. There will be, for that channel and a particular scalar measure of average distortion D between what is sent and what is received, a minimum necessary channel capacity $R(D)$.

The theorem asks, in effect, what is the 'best' channel for transmission of a message with the least possible average distortion. $R(D)$ can be defined for nonergodic information sources via a limit argument based on the ergodic decomposition of a nonergodic source into a 'sum' of ergodic sources. The RDT can be reconfigured in a control theory context, if we envision a system's topological information from the Data Rate Theorem as another form of noise, adding to the average distortion D. See Fig. 9.2.

The punctuation implied by the Data Rate Theorem (Nair et al. 2007) happens if there is a critical maximum average distortion beyond which control fails.

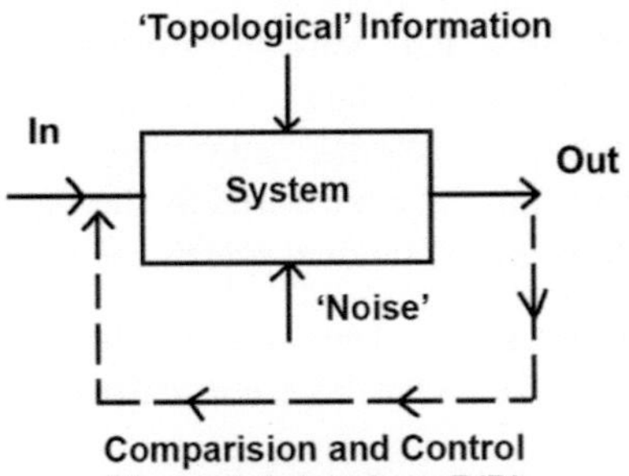

Fig. 9.2 The rate distortion control theory model. An inherently unstable system's 'topological information', according to the data rate theorem (Nair et al. 2007), is added to the effects of 'noise' to determine the minimum channel capacity $R(D)$ needed to assure transmission with average distortion D. The punctuation implied by the DRT emerges here if there is a critical maximum average distortion beyond which control fails

9.2 The Model

We provide a basic approach, applying and abducting standard formalism from statistical mechanics (Landau and Lifshitz 2007)—given a proper definition of 'temperature'—for a cognitive system.

Consider the full ensemble of high probability developmental trajectories available to the system, written as $Y_j,\ j = 1, 2, \ldots$. Each trajectory is associated with a Rate Distortion Function-defined minimum necessary channel capacity R_j for a particular maximum average scalar distortion D_j. Then, assuming some basic underlying probability model having distribution $\rho_R(C_R, x)$, where C_R is a parameter set, we can define a pseudoprobability for a meaningful 'message' Y_j sent into the system of Fig. 9.2 as

$$dP = \frac{\rho_R(C_R, R/g(Z))dR}{\int_0^\infty \rho_R(C_R, R/g(Z))dR} = \frac{\rho_R(C_R, R_j/g(Z))dR}{g(Z)} \tag{9.2}$$

since, by definition, $\rho_R(C_R, x)$ is a probability density function that integrates to one on the interval $x = 0 \rightarrow \infty$. Here, for simplicity, we only treat the continuous case. For discrete or finite-state systems, the integral is a sum and the partition function denominator becomes some more general function $h(g(Z))$.

To reiterate, Y_j is some particular trajectory, while the (generalized) integral is over all possible 'high probability' paths.

We thus, at least implicitly, impose the Shannon-McMillan Source Coding Theorem (Cover and Thomas 2006): the set of possible system trajectories can be divided in two. These are, first, a very large set of measure zero—vanishingly low probability—that is not consistent with the underlying grammar and syntax of some basic information source, and a much smaller consistent set (Khinchin 1957).

Again, R_j is the RDT channel capacity keeping average distortion less than a given limit D_j for message Y_j. $g(Z)$ is the temperature analog from above, depending on the scalar resource rate Z. $\rho(C, x)$, for physical systems, is usually taken as the Boltzmann distribution: $\rho(x) = \exp[-x]$. We suggest that, for cognitive phenomena—from the living and institutional to the machine and composite—it is necessary to move beyond analogs with physical systems. That is, it becomes necessary explore the influence of a variety of different probability distributions, including those with 'fat tails' (Derman et al. 2016; Taleb 2020), on the dynamics of cognition/regulation stability.

Again, standard methodology from statistical physics (Landau and Lifshitz 2007) identifies the denominator of Eq. (9.2) as a partition function, giving an iterated free energy-analog F as

$$\rho_F(C_F,\ F/g(Z)) \equiv \int_0^\infty \rho_R(C_R,\ R/g(Z))dR = g(Z)$$
$$g(Z) = \text{RootOf}\,(X - \rho_F(C_F,\ F/X)) \tag{9.3}$$

where 'RootOf' means setting the expression to zero and solving for X and ρ_F is an appropriate 'free energy' distribution that must be determined case-by-case for cognitive, as opposed to simple physical, phenomena. For physical phenomena, ρ_F can often be taken as the Boltzmann distribution. Again, C_X is simply an associated parameter set.

From chemical physics (Laidler 1987), we can define a cognition rate for the system as the reaction rate analog

$$L \equiv \frac{\int_{R_0}^\infty \rho_R(C_R,\ R/g(Z))dR}{\int_0^\infty \rho_R(C_R,\ R/g(Z))dR} = \frac{1}{g(Z)}\int_{R_0}^\infty \rho_R(C_R,\ R/g(Z))dR \tag{9.4}$$

R_0 is the minimum channel capacity needed to keep the average distortion below a critical value D_0 for the full system of Fig. 9.2.

The basic underlying probability model of Eq. (9.2)—via the partition function argument of Eq. (9.3)—determines system dynamics, but not system structure, and cannot be associated with a particular underlying network conformation, although these are related (Watts and Strogatz 1998; Barabasi and Albert 1999). More specifically, a set of markedly different networks may all be mapped on to any single given dynamic behavior pattern and, indeed, vice-versa, The same static network may display a spectrum of behaviors (Harush and Barzel 2017). We focus on dynamics rather than network topology, a shift from much current work.

Feynman (2000)—following Bennett (1982)—identifies 'information' as a form of free energy, using Bennett's ideal machine that turns a message directly into work. We are concerned here with an iterated, rather than a direct, construction.

F, as a free energy, is subject to symmetry-breaking transitions as $g(Z)$ varies (Pettini 2007). These symmetry changes, however, are not as associated with physical phase transitions as represented by standard group algebras. Such symmetry changes represent transitions from playing one 'game' to playing another. For example, a cognitive system may engage in foraging behaviors that trigger a predatory attack by another system. Then the game changes from 'foraging' to 'escape'.

Thus 'cognitive phase changes' involve shifts between equivalence classes of high probability developmental/behavioral pathways that are represented as *groupoids*, a generalization of the group concept such that a product is not necessarily defined for every possible element pair, although multiple products with multiple identity elements are defined (Brown 1987; Weinstein 1996).

Dynamics emerge via a first-order Onsager approximation akin to that of nonequilibrium thermodynamics (de Groot and Mazur 1984) in the gradient of an entropy measure S constructed from the 'iterated free energy' F of Eq. (9.3) by means of a

Legendre transform that should be familiar from the Black-Scholes approximation (Landau and Lifshitz 2007; Pettini 2007).

Recall from the (first-order) Onsager approximation of nonequilibrium thermodynamics that $\partial Z/\partial t \propto \partial S/\partial Z$. The full—again, standard, first-order—development is

$$\begin{aligned}
S(Z) &\equiv -F(Z) + ZdF(Z)/dZ \\
\partial Z/\partial t &\approx dS/dZ = f(Z) \\
f(Z) &= Zd^2F/dZ^2 \\
F(Z) &= \int\int \frac{f(Z)}{Z} dZdZ + C_1 Z + C_2 \\
g(Z) &= \text{RootOf}\,(X - \rho_F(C_F,\ F(Z)/X))
\end{aligned} \tag{9.5}$$

An expression for $g(Z)$ follows from the second expression of Eq. (9.3). Note that the 'diffusion coefficient' has been implicitly set equal to 1 in the second part of Eq. (9.5).

Higher order approximations are in the same realm as higher order regression models, modulo dimensional consistency. Full treatment may involve generalization of the Legendre transform via formal power series methods (Jackson et al. 2017).

There are several important points:

- Since 'RootOf' may have complex number solutions, the temperature analog $g(Z)$ enters the realm of the 'Fisher Zeros' characterizing phase transition in physical systems (e.g., Dolan et al. 2001; Fisher 1965; Ruelle 1964 Sect. 5). Such phase transitions lie at the base of the punctuated accession to consciousness in the Baars, and related, Global Workspace Models. Indeed, the 'RootOf' construction in Eq. (9.3) actually generalizes the Lambert W-function (e.g., Yi et al. 2010; Mezo and Keady 2015).
- To reiterate, information sources are not microreversible, that is, palindromes are highly improbable, for example the sequence 'ti' has far lower probability than 'it' in English. Thus, there are no 'Onsager Reciprocal Relations' in higher dimensional systems. The imposition of groupoid symmetries on cognitive phenomena appears to follow upon this directed homotopy.
- Further, there will always be a delay in the rate of provision of Z, so that, in Eq. (9.5), for example, $f(Z) = \beta - \alpha Z(t)$—an exponential model having $Z(t) = (\beta/\alpha)(1 - \exp[-\alpha t])$ —where $Z \to \beta/\alpha$ at a rate determined by α. Other dynamics are possible, such as the 'Arrhenius', $Z(t) = \beta \exp[-\alpha/t]$, with $f(Z) = (Z/\alpha)(\log(Z/\beta))^2$. However Z is constructed from the components $\mathscr{C}$, $\mathscr{H}$ and $\mathscr{M}$, so here, it is the scalar resource rate Z itself that counts.

9.3 Dispersed and Hierarchical Cognition Rates

The Basic Idea

We assume an exponential model for $dZ/dt = f(Z) = \beta - \alpha Z$ in Eq. (9.5), so that $Z(t) \to \beta/\alpha$ and

$$F = \beta Z\,(\ln(Z) - 1) - \frac{\alpha\, Z^2}{2} + C_1 Z + C_2 \tag{9.6}$$

where C_1 and C_2 are important boundary conditions.

We consider a direct, single-stage, 'Auftragstaktik' mission command system for which the basic underlying probability distribution is the Boltzmann. After some simple calculation, Eqs. (9.3) and (9.4) become

$$\begin{aligned} \exp[-F/g] &= g \\ g &= \frac{-F}{W(n,\ -F)} \\ L &= \exp[-R_0/g] \end{aligned} \tag{9.7}$$

$W(n,\ x)$ is the famous Lambert W-function of order n that solves the relation

$$W(n,\ x)\exp[W(n,\ x)] = x.$$

It is real-valued only for $n = 0,\ -1$ and $x > -\exp[-1]$.

Setting $n = 0,\ C_1 = -1,\ C_2 = 3$, we fix $R_0 = 1,\ \alpha = 1$ and let β vary as the index of 'arousal' in a Yerkes-Dodson model (Diamond et al. 2007; Wallace and Fricchione 2024) of cognition rate L in Fig. 9.3. The real-value inverse-U of the Y-D Effect is bracketed on the left by non-zero imaginary-valued component instabilities representing 'hallucination' at low, and 'panic' at high arousal.

A standard model of carcinogenesis (e.g., Belikov 2017) involves multiple triggering events that individually take place according to 'random' exponential distributions having the same rate constants. This leads to the Erlang Distribution. To show this, one can compute the moment generating function of the sum as a power of the moment generating functions of the exponentials, separately compute the moment generating function of the Erlang, and note they are the same.

Given k sequential incidents, the basic relations are

$$\begin{aligned} \rho(k,\ \lambda,\ x) &= \frac{\lambda^k x^{k-1} e^{-\lambda x}}{(k-1)!} \\ \rho(k,\ \lambda,\ F/g) &= g \end{aligned} \tag{9.8}$$

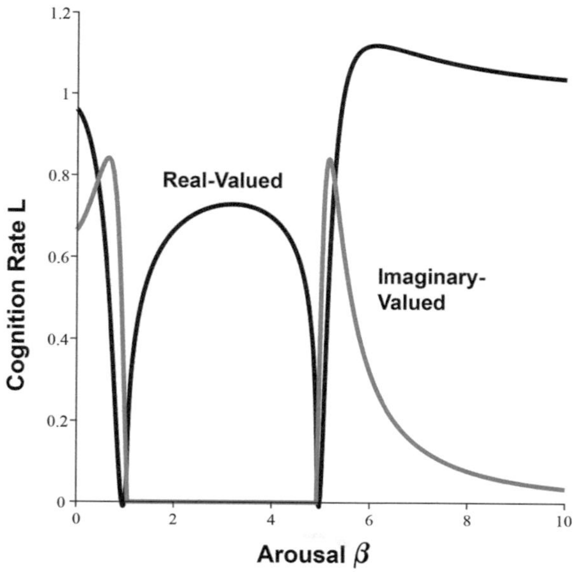

Fig. 9.3 Cognition rate versus arousal for a single-stage Auftragstaktik/mission command model follows a classic Yerkes-Dodson inverse-U pattern for the given boundary conditions. The real-value inverse-U is bracketed by nonzero imaginary-valued components on the left as 'hallucination' at low, and as 'panic' at high arousal

If, however, $k \geq 0$ is not an integer, then the denominator $(k-1)!$ is replaced by the Gamma Function $\Gamma(k)$, and the Erlang becomes the Gamma Distribution.

The relations corresponding to Eq. (9.7) for $k = 2$—a 'simple' two-step process of hierarchical cognition—are then

$$
\begin{aligned}
g &= -\frac{\lambda F}{2\, W\left(n, -\frac{\sqrt{F}}{2}\right)} \\
L &= \frac{e^{-\frac{\lambda R_0}{g}}\, (\lambda R_0 + g)}{g}
\end{aligned}
\tag{9.9}
$$

Setting $R_0 = 1$, $\lambda = 1$, $n = 0$ and imposing the same boundary conditions on F as in Fig. 9.3, we see a remarkably different pattern of cognition rate versus arousal. Only for two very narrow inverse-U spikes is the imaginary-valued component actually zero.

A 'best case' variation of the boundary condition on F, however, does lead to the coalescence of the two narrow inverse-U spikes as in Fig. 9.5. Here, $C1 = 1$, $C_2 = 33/20$ in the definition of F. Other parameters are as in Fig. 9.4. Note that, even in this example, cognition is badly stunted in comparison with the one-stage example of Fig. 9.3.

The interested reader can pursue matters to higher, or non-integer, order in k—representing added or 'smeared' levels of deliberation—where the cognitive dynamics become even more abusive.

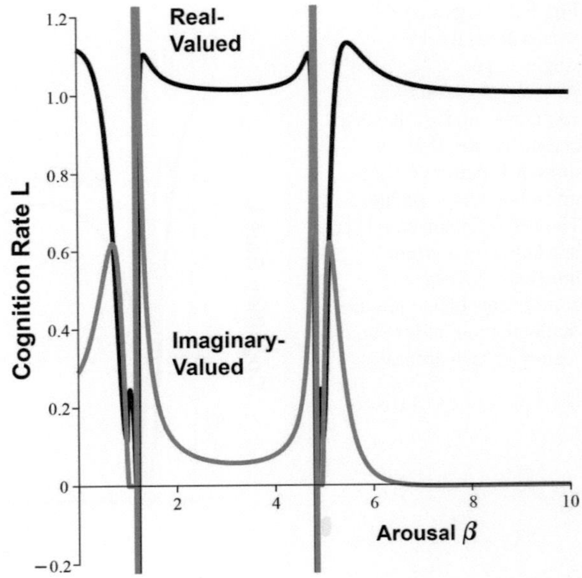

Fig. 9.4 Cognition rate versus arousal for a dual-stage hierarchical system in which two layers of authority must be traversed before a decision is reached. The boundary conditions on F are the same as in Fig. 9.3. Only for two very narrow inverse-U spikes is the imaginary-valued component zero

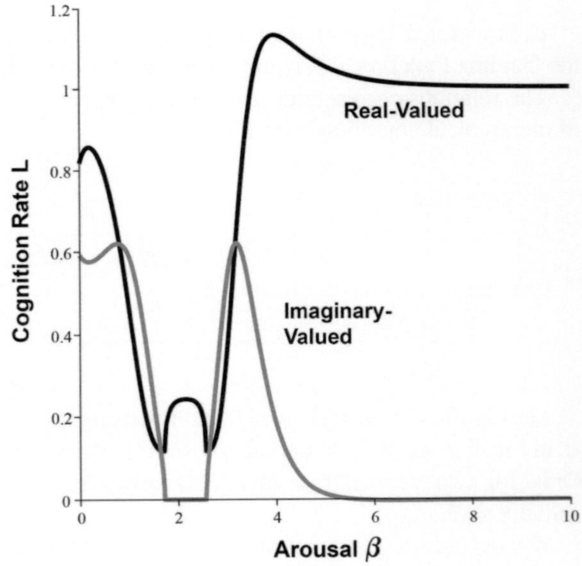

Fig. 9.5 A 'best case' treatment of the two-stage cognition process of Fig. 9.4, merging the two narrow real-valued peaks by taking $C_1 = 1,\ C_2 = 33/20$ in F. Even here, the cognition rate is badly stunted in comparison with the one-stage model of Fig. 9.3

The intermediate case in k for the Gamma Distribution is exactly solvable, after a little work, generating an analog to Fig. 9.4. Even 'some' micromanagement can be catastrophic. The relevant relations, taking $R_0 = 1$, are

$$
\begin{aligned}
g &= -\frac{\lambda F}{kW\left(n, -\frac{\left(\Gamma(k)F\lambda^{-k}\right)^{\frac{1}{k}}\lambda}{k}\right)} \\
L &= \frac{(\lambda/g)^k \exp[-\lambda/g]}{\Gamma(k+2)} \\
&\times \left(\frac{e^{\frac{\lambda}{g}}\Gamma(k+2)}{\left(\frac{\lambda}{g}\right)^k} - e^{\frac{\lambda}{2g}} \left(\frac{\lambda}{g}\right)^{-\frac{k}{2}} M_{\frac{k}{2}, \frac{k}{2}+\frac{1}{2}}\left(\frac{\lambda}{g}\right) - k - 1 \right)
\end{aligned}
\tag{9.10}
$$

where W is the Lambert W-function and M the Whittaker M-function.

The Mathematical Appendix reconsiders the Boltzmann versus second-order Erlang model in discrete, rather than continuous circumstances, when there are only two possible states, $R_{\pm} = R_0 \pm \delta$, $\delta \ll R_0$. R_+ is taken as the cognitively 'on' state.

In addition, the Appendix reexamines the 'best case' of Fig. 9.5 for the Gamma Distribution, assuming variation in k.

Stochastic Extension

Under stochastic burden, it is possible to derive F from the assumption of stability imposed by evolutionary selection pressure on some particular function $Q(Z)$, i.e., first assuming $< dQ_t >= 0$ and then calculating F from the relation

$$
dZ_t = (Z_t d^2F/dZ^2)dt + \sigma h(Z_t)dB_t \tag{9.11}
$$

via the Ito Chain Rule applied to Q. Then, after some development,

$$
F(Z) = \int \left(\int -\frac{\sigma^2 h(Z)^2 \left(\frac{d^2}{dZ^2}Q(Z)\right)}{2Z\left(\frac{d}{dZ}Q(Z)\right)} dZ \right) dZ + C_1 Z + C_2 \tag{9.12}
$$

with g and L determined from the basic underlying probability distribution as in Eqs. (9.7) and (9.9).

We make one detailed application, assuming that institutional signal detection follows something like the classic Weber-Fechner Law of sensory perception (Reina et al. 2018), that is, 'perceived sensation' is scaled as the logarithm of the of the signal energy, in 'decibels' of incoming and internal signal strength.

We are then interested in the nonequilibrium steady state for $Q(Z) \equiv \ln(1+Z)$, calculating $< d\ln(1+Z_t) >= 0$ for ordinary volatility, $h(Z) = Z$, in Eq. (9.12). Then

$$
F(Z) = \frac{\sigma^2 \left(\frac{Z^2}{2} - \ln(1+Z)(1+Z) + 1 + Z\right)}{2} + C_1 Z + C_2 \tag{9.13}
$$

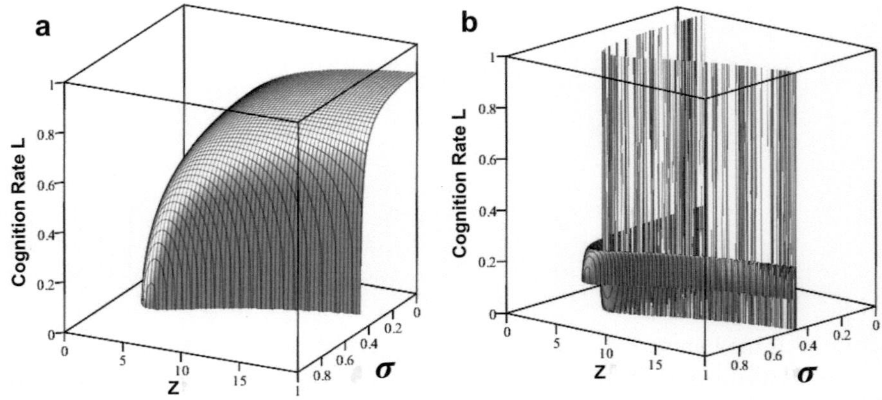

Fig. 9.6 Joint variation of cognition rate with σ and Z for **a** Boltzmann and **b** Second-order Erlang distributions. As in Figs. 9.3 and 9.4, while the one-stage Boltzmann distribution shows an even progression from the 'easy' to the 'hard' Yerkes-Dodson pattern with increasing 'noise' σ, the two-stage hierarchy displays severely stunted cognitive dynamics

We again take $C_1 = -1$, $C_2 = 3$, $n = 0$ and calculate the cognition rates for varying σ and Z for the single-stage Boltzmann and two-stage Erlang distributions in Fig. 9.6. As in Figs. 9.3 and 9.4, while the Boltzmann distribution, Fig. 9.6a, shows an even progression from the 'easy' to the 'hard' Yerkes-Dodson pattern with increasing 'noise' σ, the two-fold hierarchy of Fig. 9.6b displays stunted cognitive dynamics. Both, however, show collapse of cognition under sufficient noise burden.

Taking a Pieron's Law perspective (Reina et al. 2018)—that the system detects signals at a rate $\propto Z^{-m}$, $m > 0$—then the nonequilibrium steady state condition $< dZ_t^{-m} >= 0$, again for ordinary volatility under the Ito Chain Rule, leads to

$$F(Z) = \frac{(m+1)\sigma^2 Z^2}{4} + C_1 Z + C_2 \tag{9.14}$$

Setting $m = 2$, calculations of cognition rate for the two distributions essentially reprises Fig. 9.6.

Time-and-Resource Constraints

More typically, one usually has a large, multicomponent, interlinked structure of $i = 1, \ldots, n$ subcomponents that must act together under constraints of both resource rates Z_i and available length of time T_i under which a component must act. The resource and time constraints are $Z = \sum_i Z_i$ and $T = \sum_i T_i$, and one seeks to optimize some collective performance index $\sum_i Q_i(Z_i, T_i)$ under those constraints, leading to the Lagrangian system of equations

$$
\begin{aligned}
\mathcal{L} &= \sum_i Q_i(Z_i, T_i) + \lambda \left(Z - \sum_i Z_i \right) + \mu \left(T - \sum_i T_i \right) \\
\partial \mathcal{L}/\partial Z_i &= \partial Q_i/\partial Z_i - \lambda = 0 \\
\partial \mathcal{L}/\partial T_i &= \partial Q_i/\partial T_i - \mu = 0 \\
\partial \mathcal{L}/\partial Z &= \lambda,\ \partial \mathcal{L}/\partial T = \mu
\end{aligned}
\tag{9.15}
$$

where λ and μ are interpreted as *environmentally-imposed shadow price constraints* in the economic sense (e.g., Jin et al. 2008; Robinson 1993).

Mild abuse of formalism from the last line of Eq. (9.15) leads to $dZ/dT \sim \mu/\lambda \equiv \omega$ for the *total* resource rate and the *total* available time. This permits definition of an underlying base differential equation for the full multicomponent system as

$$
\begin{aligned}
dZ/dt \sim \omega &= F d^2 F/dZ^2 \\
F(Z) &= \omega Z\,(\ln(Z) - 1) + C_1 Z + C_2
\end{aligned}
\tag{9.16}
$$

This characterization of F, as above, permits calculation of the cognition rates for the Boltzmann and Erlang order 2 distributions under the conditions of Figs. 9.3, 9.4, and 9.6, leading to Fig. 9.7. Somewhat surprisingly, there is a bifurcation according to the sign of the shadow price ratio, with the single-stage Boltzmann distribution still far more regular than the two-stage Erlang system, which again seems stunted and incoherent.

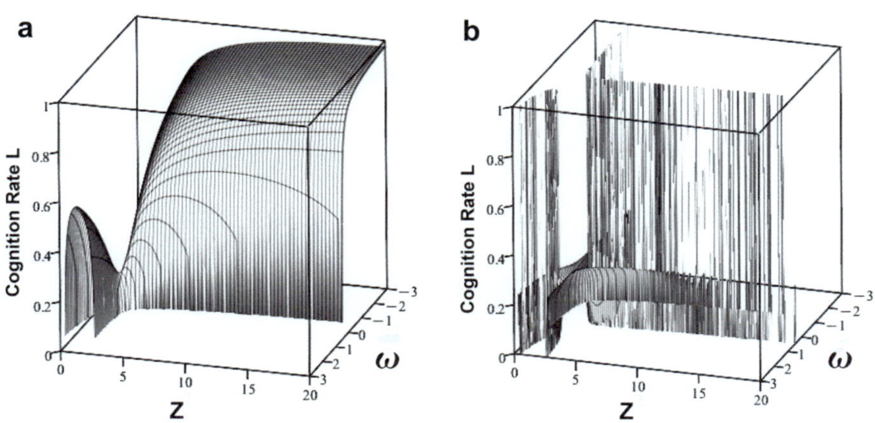

Fig. 9.7 Analog to Fig. 9.6 for the Lagrange optimization under time and resource rate constraints. The figure displays the joint variation of cognition rate with shadow price ratio ω and total resource rate Z. Here, as above, $C_1 = -1$, $C_2 = 3$, likewise assigning Eqs. (9.7) and (9.9). There is a bifurcation according to the sign of the shadow price ratio, with the single-stage Boltzmann distribution **a** more regular than the two-stage Erlang system **b**, which, again, is stunted and incoherent

It is possible, if algebraically exasperating, to iterate the argument a further step, exploring the effect of 'noise' via the SDE

$$dZ_t = \omega dt + \sigma Z_t dB_t \tag{9.17}$$

imposing, under Eq. (9.16) and the Ito Chain Rule, the nonequilibrium steady state condition $< dL_t >= 0$ on the single and double-stage expressions for cognition rate from Eqs. (9.7) and (9.9). Solving for $\sigma^2(Z, \omega)$, one again finds coherence for the one-stage Boltzmann, and incoherence for the two-stage Erlang model. Details are left as an exercise.

9.4 The Dynamics of Delay

We have, in this one- versus two-level analysis, implicitly assumed a best-case model for which there is no significant added time lag in the ability to respond to environmental perturbations, i.e., no 'OODA Loop burden' (Osinga 2007). As a general matter, however, for all examples, imposing higher order Erlang distributions—adding more supervisory levels—enters the fatal realm that follows from having an opponent 'get inside your command cycle'.

Even the most elementary one-stage model can be driven into instability by sufficient delay. Consider an 'exponential' system

$$\begin{aligned} dZ/dt &= \beta - \alpha Z(t) \\ Z(t) &= \frac{\beta}{\alpha}(1 - \exp[-\alpha t]) \rightarrow \frac{\beta}{\alpha} \end{aligned} \tag{9.18}$$

Imposing a fixed delay, say τ, so that $dZ/dt = \beta - \alpha Z(t - \tau)$, leads to the relations

$$\begin{aligned} Z(t) &= \frac{\beta}{\alpha}(1 - \exp[St]) \\ S &= \frac{W(n, -\alpha\tau)}{\tau} \\ dZ/dt &= S\left(Z - \frac{\beta}{\alpha}\right) = Zd^2F/dZ^2 \\ F(Z) &= \frac{S\,Z^2}{2} - \frac{S\beta\,(Z\ln(Z) - Z)}{\alpha} + C_1 Z + C_2 \\ dZ_t &= S\left(Z_t - \frac{\beta}{\alpha}\right)dt + \sigma Z_t dB_t \end{aligned} \tag{9.19}$$

where, again, $W(n, X)$ is the Lambert W-function of orders $n = 0, -1$, real-valued only for $X \geq -\exp[-1]$. Sufficiently large $\alpha\tau$, in the expression for S, will drive even this system into thrashing instability. If the imaginary component of S is zero, the real component is always negative, leading to stochastic instabilities with rising σ in the stochastic differential equation calculations.

Figure 9.8 reprises Figs. 9.3 and 9.4, the one-step and two-step models, under fixed delays, taking $n = 0$, $\alpha = 1$, $C_1 = -1$, $C_2 = 3$. Note the imaginary-onset disjunctions at $\tau = \exp[-1]$ for both models. The two-step system is again stunted and incoherent compared to the single step.

Assuming first and second-order Erlang Distributions, even simple distributed delay relations such as

$$
\begin{aligned}
dZ/dt &= \beta - \alpha \int_0^t Z(t-\tau)\lambda \exp[-\lambda\tau] d\tau \\
dZ/dt &= \beta - \alpha \int_0^t Z(t-\tau)\lambda^2 \tau \exp[-\lambda\tau] d\tau
\end{aligned}
\tag{9.20}
$$

which can be directly solved via the Laplace Transform method, are found to drive thrashing instabilities, depending on complicated interactions between α and λ.

Higher order Erlang systems will, by their sequential staging, suffer progressively longer delays than lower, since, from Eq. (9.8), $< \tau >= k/\lambda$.

We can carry the discussion of delay a step further, again invoking Pieron's Law of system response time, that $T \propto Z^{-m}$ for $m > 0$ (Reina et al. 2018). The last relation of Eq. (9.19) allows calculation of the nonequilibrium steady state $< dZ_t^{-m} >= 0$ according to the Ito Chain Rule, leading to

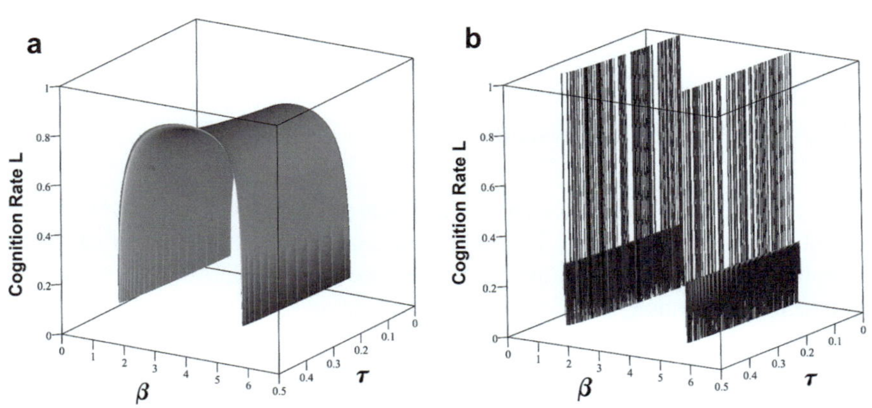

Fig. 9.8 Reprising the models of Figs. 9.3 and 9.4 under fixed delay. **a** The one-step model under the Boltzmann distribution. **b** The two-step Erlang distribution system. Here, $n = 0$, $\alpha = 1$, $C_1 = -1$, $C_2 = 3$. Note the non-zero imaginary part disjunctions in both cases at $\tau = \exp[-1]$. The two-step system is again stunted and incoherent compared to the single step

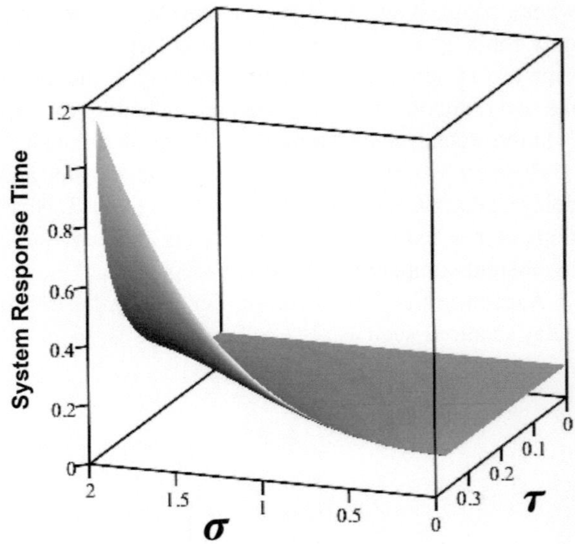

Fig. 9.9 Response time $T = Z^{-2}$, $\alpha = 1$, $\beta = 3$. For $S(n, \alpha, \tau)$, $n = -1$, $\alpha = 1$. T increases synergistically with σ and τ until the critical point $\alpha\tau = \exp[-1] \approx 0.37$, after which instability becomes manifest

$$
\begin{aligned}
Z_{nss} &= \frac{S(n, \alpha, \tau)(\beta/\alpha)}{S(n, \alpha, \tau) - \frac{(m+1)\sigma^2}{2}} \\
T_{nss} &\sim Z_{nss}^{-m}
\end{aligned}
\tag{9.21}
$$

For Fig. 9.9, $T = Z^{-2}$, $\alpha = 1$, $\beta = 3$ and for S, $n = -1$, $\alpha = 1$. T increases synergistically with both σ and τ until the critical point $\alpha\tau = \exp[-1] \approx 0.37$, where thrashing sets in.

9.5 A Multiplicative System

The Erlang/Gamma Distribution applies to a sequential series of independent, exponentially distributed, 'events'. It is, perhaps, conceivable that events might have multiplicative rather than additive influence, like an exploding infection or other reproducing population, or cancer metastasis, leading to a Lognormal Distribution model:

$$\rho(x, \mu, \sigma) = \frac{\exp[-(\ln(x) - \mu)^2/(2\sigma^2)]}{x\sigma\sqrt{2\pi}}$$
$$g_{\pm} = \frac{F}{e^{\mu \pm \sigma\sqrt{-\ln(2\pi F^2\sigma^2)}}}$$
$$L = \frac{\text{erf}\left(\frac{\sqrt{2}(\ln(g)-\ln(R_0)+\mu)}{2\sigma}\right)}{2} + \frac{1}{2} \quad (9.22)$$

We impose an exponential model on dZ/dt, generating the expression for F of Eq. (9.6). Here, $\alpha = 1$, $C_1 = -1$, $C_2 = 3$, and β is again the index of 'arousal'.

Figure 9.10 shows the corresponding real- and imaginary-valued cognition rates $L(\beta)$ for g_-, setting $R_0 = \mu = \sigma = 1$. This recovers an analog to the sequential additive model of Fig. 9.4, with only two, very narrow, real-valued spikes. The rest follows.

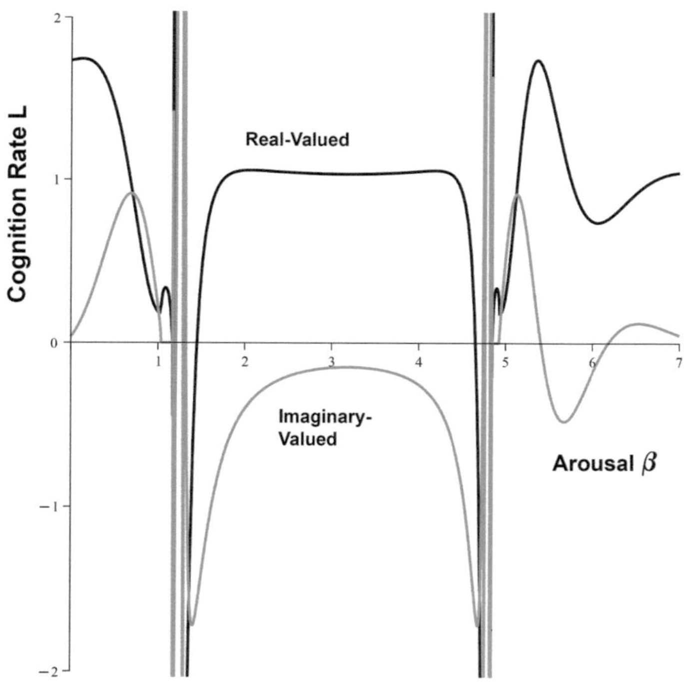

Fig. 9.10 Analog to Fig. 9.4 for the multiplicative model of Eq. (9.22), using g_-. $R_0 = \mu = \sigma = 1$, $C_1 = -1$, $C_2 = 3$. There are only two narrow stable cognition rate spikes

9.6 A Crack-Propagation Model

Crack-propagation system failure models are often based on the Weibull Distribution leading, here, to the relations

$$\begin{aligned}\rho(k,\,\lambda,\,x) &= \frac{k\left(\frac{x}{\lambda}\right)^{k-1}\mathrm{e}^{-\left(\frac{x}{\lambda}\right)^{k}}}{\lambda}\\ g &= \frac{F\left(\frac{F}{k}\right)^{-\frac{1}{k}}\left(-\frac{F}{kW\left(n,-\frac{F}{k}\right)}\right)^{\frac{1}{k}}}{\lambda}\\ L &= \mathrm{e}^{-\left(\frac{R_0}{g\lambda}\right)^{k}}\end{aligned} \tag{9.23}$$

Taking F as in Eq. (9.6), with $n = 0,\,-1,\,R_0 = 1,\,C_1 = -1,\,C_2 = 3$, leads to the starkly fragmented cognitive dynamics of Fig. 9.11. Again, there are only two narrow stable cognition spikes with rising arousal β.

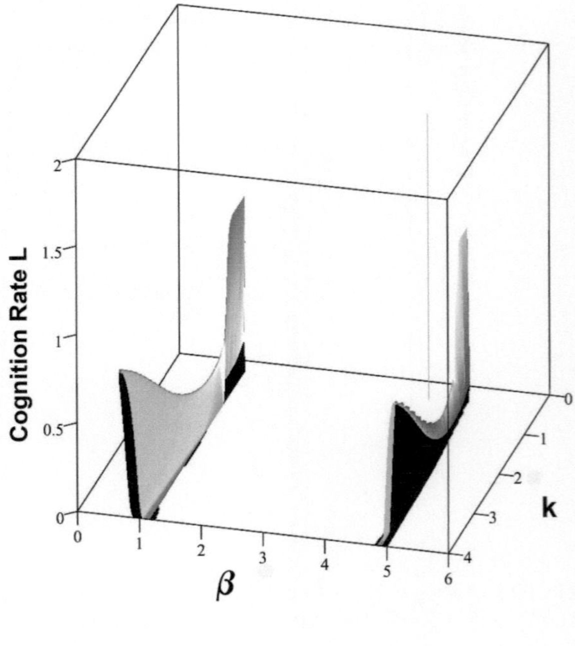

Fig. 9.11 Stable cognitive dynamics structure for a Weibull distribution crack-propagation model, $n = 0,\,-1,\,R_0 = 1$ under the conditions of Eq. (9.6), $C_1 = -1,\,C_2 = 3$. Again, there are only two narrow stable cognition spikes

9.7 Determining the Underlying Probability Distribution

Predicting the dynamics of system cognition depends, in this model, on identification of the foundational probability distribution—or the spectrum of distributions under different forms of stress—characterizing the system under study. The most direct method might be through empirical observation of the system's 'hazard rate', the form of the function $Q(x)$ representing the rate at which it detects a new signal, given a signal rate x, for different signal classes.

For a distribution $\rho(x)$ with associated hazard rate $Q(x)$, the basic relations are

$$\begin{aligned} Q(x) &\equiv \frac{\rho(x)}{1-\int_0^x \rho(u)du} \\ \rho(x) &= \lim_{\varepsilon \to 0} Q(\varepsilon)\, e^{\int_\varepsilon^x \left(-Q(Z)+\frac{\frac{d}{dZ}Q(Z)}{Q(Z)}\right)dZ} \end{aligned} \tag{9.24}$$

For the exponential distribution $\rho(x) = \lambda \exp[-\lambda x]$, $Q(x) = \lambda$.

For the Gamma distribution in arbitrary k, λ,

$$Q(k, \lambda, x) = -\frac{x^{k-1}\lambda^k k\,(k+1)}{-e^{\lambda x}\Gamma(k+2) + \left((\lambda x)^{-\frac{k}{2}} M_{\frac{k}{2},\frac{k}{2}+\frac{1}{2}}(\lambda x)\, e^{\frac{\lambda x}{2}} + k + 1\right) x^k \lambda^k} \tag{9.25}$$

where M is the Whittaker M Function of appropriate order, recalling that the Erlang Distribution is defined for integer k.

For the multiplicative system Lognormal Distribution,

$$Q(\mu, \sigma, x) = -\frac{e^{-\frac{(\ln(x)-\mu)^2}{2\sigma^2}}\sqrt{2}}{x\sigma\sqrt{\pi}\left(-1+\operatorname{erf}\left(\frac{\sqrt{2}\,(\ln(x)-\mu)}{2\sigma}\right)\right)} \tag{9.26}$$

where erf is the Error Function.

For the 'crack propagation' Weibull Distribution,

$$Q(k, \lambda, x) = k\lambda^{-k}x^{k-1} \tag{9.27}$$

Figure 9.12 shows the form of these relations for the Gamma (including the Boltzmann) and the Lognormal Distributions. For the Gamma, $\lambda = 1$, and for the Lognormal, $\mu = \sigma = 1$.

Assuming a simple monotonically declining hazard rate $Q(\lambda, x) = \lambda/(\lambda x + 1)$ implies $\rho(\lambda, x) = \lambda/(\lambda x + 1)^2$, a 'fat-tailed' distribution without finite first or second moments. Calculating 'temperature' g and cognition rate L as above, then imposing the exponential model of Eq. (9.6) on F as in Fig. 9.3 leads—somewhat

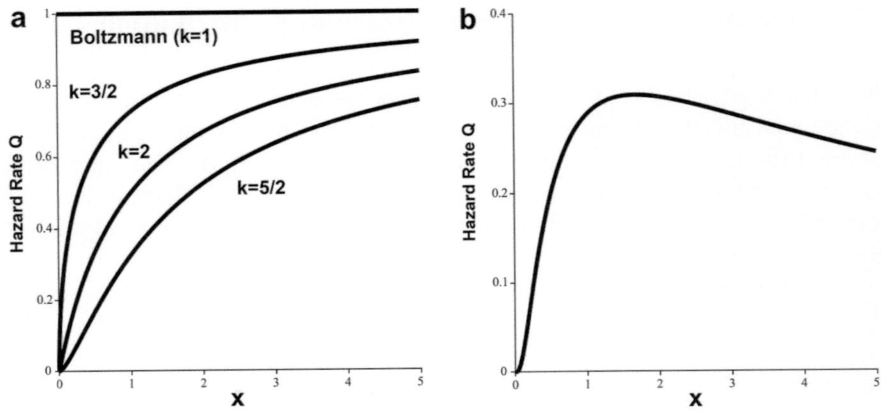

Fig. 9.12 Hazard rate forms. **a** For the Gamma distribution at various k, setting $\lambda = 1$, including the Boltzmann distribution, k = 1. **b** For the Lognormal distribution, $\mu = \sigma = 1$

surprisingly—to a close analog to Fig. 9.3 for cognition rate versus arousal. Details are left as an exercise.

As shown in the Mathematical Appendix, imposing a U-shaped hazard function, using a generalized Gompertz Distribution (El-Gohary et al. 2013), via numerical methods, leads to severely constrained cognition, replicating the two narrow real-valued peaks of Fig. 9.4.

By contrast, an 'ordinary' Gompertz Distribution, with an exponentially-increasing hazard rate, can also be addressed by numerical methods and generates the inverse-U of Fig. 9.3.

In sum, empirical observation of a cognitive system's hazard rate or rates under varying signal strengths and modes permits inference of the underlying probability distribution (or the spectrum of distributions under different excitations), and hence inference of cognitive dynamics and dysfunction under differing selection pressures.

Further, since the form of the underlying probability distribution has profound impact on the dynamics of cognition, it is likely an object of evolutionary process, leading to 'distribution tuning' under stress.

9.8 Discussion

Those of us accustomed to acting within 'normal' bureaucracies will, perhaps, recognize something of the degradation, distortion, and stunting imposed by increasing levels of 'decision structure' under the critical time constraints so fondly exploited by John Boyd and his disciples.

The striking difficulties encountered by rigidly-controlled Russian forces in the Ukraine are well known and summarized in the Appendix from this multi-stage perspective by the Perplexity AI system.

Indeed, the term 'workplace autonomy' (Lerner, 2024) is often cited in civilian enterprise as a central tool for ensuring both high productivity and retention of employees.

What this work suggests, among other matters, is that the emergence of an Erlang/Gamma Distribution of order above one in an institutional pattern of times-to-decision is very much not a good thing.

In addition, the probability models explored here, based firmly on the asymptotic limit theorems of information and control theories, might, with some effort, be converted into robust statistical tools for the study and remediation of pathologies of cognition at and across a considerable spectrum of scales and levels of organization not confined to the study of organized conflict.

Unfortunately, as the works of the master statisticians of the early 20th Century demonstrate, the conversion of asymptotic limit theorems into robust, verifiable statistical tools is a fraught enterprise. First, the limit theorems associated with 'ordinary' statistics are, themselves recondite, as those who have forged through the proofs of the Central Limit, Renewal, Martingale, and related Theorems have found. The Theorems that characterize and constrain information and control theories are no less deep and subtle. In addition, the first-order 'statistical mechanics' and 'nonequilibrium thermodynamics' methods used to derive the probability models developed here are simplistic—indeed entry-level—approximations abducted from physical theory. Analysis of real-world, real-time data would be needed to map out the necessary extensions and modifications of both theory and the tools that follow from that theory. This was, indeed, much the experience in the emergence of 'ordinary' parametric and nonparametric statistical methodology.

In sum, the mathematical structures invoked here, complicated as they may seem, provide only the most limited initial starting point. After all, as Maturana and Varela (1980) assert, cognition lies at the very heart of biological mystery. A full-scale address, it seems, requires even deeper formal approaches than explored here, such as phase transition driven by nonergodic groupoid symmetry-breaking and associated arguments (e.g., Wallace 2022, 2023).

As reviewers of this chapter have noted, we do not provide empirical validation of model outputs such as cognitive transitions or collapse thresholds. One put the matter this way,

> ...[A] supervised learning approach is proposed in which the target variable corresponds to one of the latent quantities that cannot be directly observed but are essential to the system's dynamics, such as decision delay (τ) or the level of operational distortion (σ). These parameters, which are crucial for determining system stability and transitions toward dysfunctional cognitive states, can be estimated from historical or simulated data using regression models.
>
> The observable features used for learning might include the average duration of decisions, the number of hierarchical levels involved, the frequency of communication between operational units, the variability of decision outcomes, or the amount of resources used. These variables, derived from operational logs or synthetic datasets, would serve to train a predictive model capable of inferring the value of the theoretical parameter selected as the target.

At present, the author simply does not have access to the resources needed for such studies. A clever reader, however, can take this material and run with it.

Appendix

Two-State Models

Here, we assume there are only two possible states, 'on' with $R_+ = R_0 + \delta,\ \delta \ll R_0$, and 'off', $R_- = R_0 - \delta$, again under the Boltzmann and two-stage Erlang probability distributions. We set $R_0 = 1$. Under these distributions, the essential relations for the partition functions and cognition rates become

$$
\begin{aligned}
e^{-\frac{F}{g}} &= e^{-\frac{1+\delta}{g}} + e^{-\frac{1-\delta}{g}} \\
\frac{F\, e^{-\frac{F}{g}}}{g} &= \frac{(1+\delta)\, e^{-\frac{1+\delta}{g}}}{g} + \frac{(1-\delta)\, e^{-\frac{1-\delta}{g}}}{g}
\end{aligned}
\tag{9.28}
$$

$$
\begin{aligned}
L &= \frac{e^{-\frac{1+\delta}{g}}}{e^{-\frac{1+\delta}{g}} + e^{-\frac{1-\delta}{g}}} \\
L &= \frac{(1+\delta)\, e^{-\frac{1+\delta}{g}}}{g\left(\frac{(1+\delta)e^{-\frac{1+\delta}{g}}}{g} + \frac{(1-\delta)e^{-\frac{1-\delta}{g}}}{g}\right)}
\end{aligned}
\tag{9.29}
$$

Equation(9.28) can be expanded as a series in δ, giving first-order approximations for g as

$$
\begin{aligned}
g &\approx \frac{1-F}{\ln(2)} + \mathcal{O}(\delta^2) \\
g &\approx \frac{1-F}{\ln(2/F)} + \mathcal{O}(\delta^2)
\end{aligned}
\tag{9.30}
$$

We drop the terms of order δ^2.

F is taken as in Eq. (9.6), setting $\alpha = 1,\ C_1 = -1,\ C_2 = 3$, with β the index of arousal. This generates the cognition rate graphs of Fig. 9.13a for the Boltzmann and 13b for the two-stage Erlang distribution. The two-stage model is bifurcated and, again, grossly distorted, in comparison with the Boltzmann example.

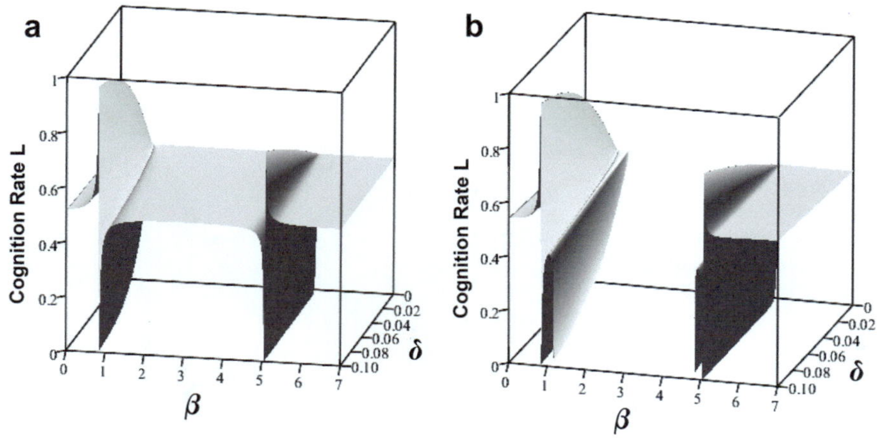

Fig. 9.13 F is as in Eq. (9.6), $\alpha = 1$, $C_1 = -1$, $C_2 = 3$, with β the index of arousal. **a** Cognition rate for the Boltzmann distribution. **b** Cognition rate for the two-stage Erlang distribution. Again, the two-stage model is grossly distorted, in comparison with the Boltzmann example. Two-state examples based on the Lognormal distribution are recognizably similar to the two-stage Erlang case

For the two-stage Lognormal Distribution, again setting $R_0 = 1$,

$$
\begin{aligned}
g &\approx \sqrt{F}\, 2^{\frac{\sigma^2}{\ln(F)}} \mathrm{e}^{\sigma^2-\mu} + \mathcal{O}(\delta^2) \\
L &= \frac{g\sqrt{2}\,\mathrm{e}^{-\frac{\left(\ln\left(\frac{1+\delta}{g}\right)-\mu\right)^2}{2\sigma^2}}}{2\,(1+\delta)\,\sigma\sqrt{\pi}\left(\frac{g\sqrt{2}\,\mathrm{e}^{-\frac{\left(\ln\left(\frac{1+\delta}{g}\right)-\mu\right)^2}{2\sigma^2}}}{2(1+\delta)\sigma\sqrt{\pi}} + \frac{g\sqrt{2}\,\mathrm{e}^{-\frac{\left(\ln\left(\frac{1-\delta}{g}\right)-\mu\right)^2}{2\sigma^2}}}{2(1-\delta)\sigma\sqrt{\pi}}\right)}
\end{aligned}
\tag{9.31}
$$

The algebraic expression for g is different, but the graph of $L(\delta, \beta)$ is recognizably similar to the two-stage Erlang case.

For the Weibull Distribution crack propagation model, setting $\lambda = R_0 = 1$,

$$
\begin{aligned}
g &\approx \frac{1-F}{\ln\left(2^{\frac{1}{k}} F^{-\frac{k-1}{k}}\right)} + \mathcal{O}(\delta^2) \\
L &= \frac{k\left(\frac{\delta+1}{g}\right)^{k-1} \mathrm{e}^{-\left(\frac{\delta+1}{g}\right)^k}}{k\left(\frac{\delta+1}{g}\right)^{k-1} \mathrm{e}^{-\left(\frac{\delta+1}{g}\right)^k} + k\left(\frac{1-\delta}{g}\right)^{k-1} \mathrm{e}^{-\left(\frac{1-\delta}{g}\right)^k}}
\end{aligned}
\tag{9.32}
$$

The expression for g is analogous to those for the Boltzmann and Erlang Distributions. Fixing $\delta \ll 1$, the graph of $L(k, \beta)$, however, is singularly irregular and very highly bifurcated.

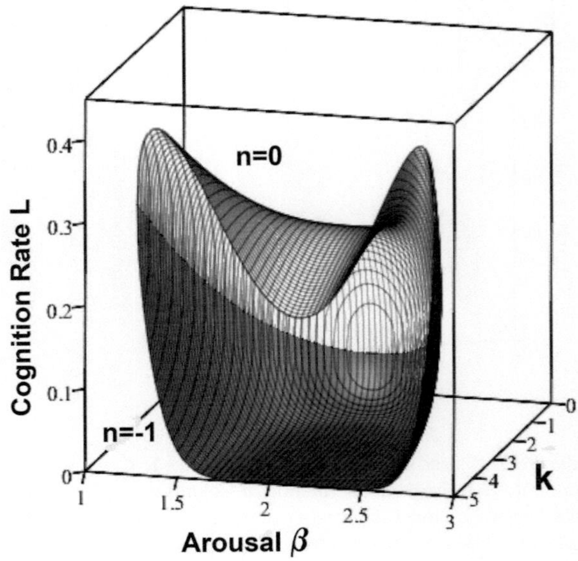

Fig. 9.14 Extension of the 'best case' from Fig. 9.5 to the Gamma Distribution model with variable k, the depth-of-command. The solution is bifurcated according to the order of the Lambert W-function defining g. Multi-layer decision structures remain cognitively stunted in this model

'Best Case' Analysis for the Gamma Distribution

Using Eq. (9.10), we can 'easily' recapitulate the analysis leading to Fig. 9.5, again, in the definition for F setting $C_1 = 1,\ C_2 = 33/20$ and taking $R_0 = 1$. The result is shown in Fig. 9.14. The result is bifurcated, with the upper section taking $n = 0$ in the Lambert W-function, and the lower corresponding to $n = -1$. Multi-layer decision structures, indexed by k, remain cognitively stunted in this model.

U-Shaped Hazard Function

The generalized Gompertz Distribution (El-Gohary et al. 2013) is characterized by the relations

$$\rho(\lambda,\, c,\, \theta,\, x) = \theta\lambda\, e^{cx} e^{-\frac{\lambda(e^{cx}-1)}{c}} \left(1 - e^{-\frac{\lambda(e^{cx}-1)}{c}}\right)^{\theta-1}$$

$$Q = \frac{\theta\lambda\, e^{cx-\frac{\lambda(e^{cx}-1)}{c}} \left(1 - e^{-\frac{\lambda(e^{cx}-1)}{c}}\right)^{\theta-1}}{1 - \left(1 - e^{-\frac{\lambda(e^{cx}-1)}{c}}\right)^{\theta}}$$

$$\theta\lambda\, e^{\frac{cF}{g}} e^{-\frac{\lambda\left(e^{\frac{cF}{g}}-1\right)}{c}} \left(1 - e^{-\frac{\lambda\left(e^{\frac{cF}{g}}-1\right)}{c}}\right)^{\theta-1} = g$$

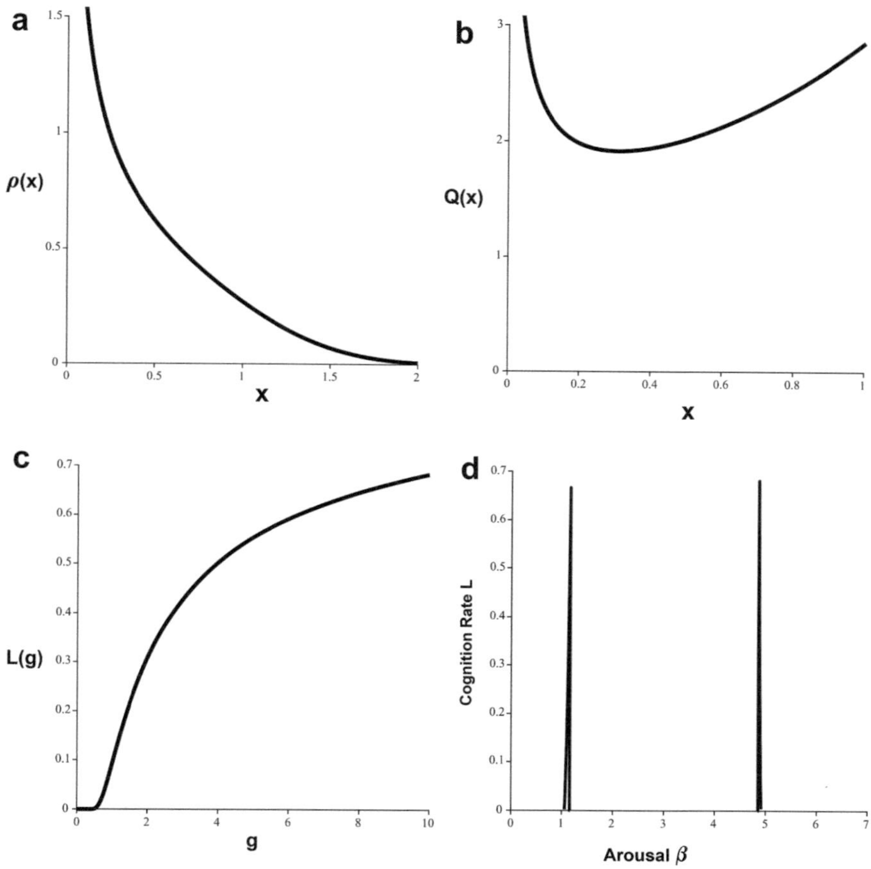

Fig. 9.15 Cognitive dynamics for the generalized Gompertz distribution, setting $\lambda = c = 1$, $\theta = 1/2$, $R_0 = 1$, under Eq. (9.6). The relation between F and g has been solved numerically. **a** $\rho(x)$. **b** Hazard rate. **c** Cognition rate as a function of g. **d** Cognition rate versus 'arousal', replicating Fig. 9.4

$$L = 1 - \left(1 - e^{-\frac{\lambda\left(e^{\frac{cR_0}{g}} - 1\right)}{c}}\right)^{\theta} \tag{9.33}$$

Setting $\lambda = c = 1$, $\theta = 1/2$, $R_0 = 1$ and imposing Eq. (9.6) with $\alpha = 1$, $C_1 = -1$, $C_2 = 3$ allows numerical solution to the third expression of Eq. (9.33) for g in terms of the arousal β, leading to the results of Fig. 9.15. Here, a U-shaped hazard rate replicates the two narrow cognition rate peaks of Fig. 9.4.

Russia and The Ukraine

As compiled by Perplexity AI, 2/21/2025.

Recent research on institutional cognition…suggests that hierarchical, detailed command structures can be modeled using a multi-stage cancer model based on the Erlang distribution [1]. This perspective provides an interesting lens through which to analyze the relative failure of the Russian military effort in Ukraine.

The Erlang distribution describes the probability of multiple independent random events occurring by a given time [2]. In the context of institutional cognition and military command, these 'events' can be interpreted as decision-making stages or information transmission steps within a hierarchical structure.

Russian Command Structure and Its Failures

The Russian military's command structure in Ukraine exhibited several characteristics that align with this multi-stage model:

1. **Hierarchical and Centralized Decision-Making**: The Russian forces relied heavily on a top-down approach, with detailed orders coming from higher echelons [4]. This aligns with a multi-stage model where each decision or order must pass through multiple layers before reaching the tactical level.
2. **Lack of Initiative at Lower Levels**: Russian units often stopped and waited for instructions from senior commanders when faced with unforeseen circumstances [6]. This behavior is consistent with a system where multiple 'events' (approvals or decisions) must occur before action can be taken.
3. **Delayed Response to Changing Situations**: The Russian military demonstrated poor adaptability to rapidly changing battlefield conditions [6]. In the Erlang distribution model, this could be interpreted as a system requiring too many stages (events) to complete before a response can be formulated.
4. **Vulnerability of Command Posts**: Ukrainian forces successfully targeted Russian command posts multiple times, significantly degrading their ability to plan and conduct coordinated operations [5]. This vulnerability is exacerbated in a system that relies heavily on centralized command, as modeled by the multi-stage approach.

Contrast with Mission Command

The Ukrainian military, on the other hand, adopted a more flexible approach aligned with modern mission command principles:

1. **Decentralized Decision-Making**: Ukrainian forces demonstrated a capacity for mission-oriented command, allowing for more autonomous operations at lower levels [4].
2. **Adaptability**: The Ukrainian military showed greater ability to adapt to changing situations, consistent with a system requiring fewer 'stages' or decision points.
3. **Resilience**: By distributing decision-making authority, the Ukrainian command structure was more resilient to disruptions, unlike the vulnerable multi-stage Russian system.

Implications

Viewing the Russian command failures through the lens of …[a]…multi-stage model highlights how a hierarchical, detailed command structure can become a liability in modern warfare. Each stage in the decision-making process introduces potential delays and vulnerabilities, much like how multiple stages in the Erlang distribution model of cancer development increase the waiting time for an event to occur [3].

This perspective underscores the advantages of mission command, which reduces the number of 'stages' required for decision-making and action, thereby increasing responsiveness and resilience on the battlefield. The Russian experience in Ukraine serves as a stark illustration of how outdated command structures can significantly hamper military effectiveness in contemporary conflicts.

Citations

1. https://www.mdpi.com/1099-4300/24/8/1070.
2. https://peerj.com/articles/11976/.
3. https://www.nature.com/articles/s41598-017-12448-7.
4. https://www.defensa.gob.es/documents/2073105/2320887/la_guerra_de_ucrania_cuando_aprender_es_lo_importante_2025_dieeeo07_eng.pdf.
5. https://www.armyupress.army.mil/Journals/Military-Review/English-Edition-Archives/May-June-2023/Graveyard-of-Command-Posts/.
6. https://indsr.org.tw/uploads/indsr/files/202311/7eba0534-512b-4275-a2f5-9d7b3ef20e1d.pdf.

References

Atlan, H., I. Cohen, 1998, Immune information, self-organization, and meaning, International Immunology, 10:711–717.

Barabasi, A., R. Albert, 1999, Emergence of scaling in random networks, Science 286:509–512.

Belikov, A., 2017, The number of key carcinogenic events can be predicted from cancer incidence, Scientific Reports, 7: 12170, https://doi.org/10.1038/s41598-017-12448-7.

Bennett, C.H., 1982, The thermodynamics of computation, International Journal of Theoretical Physics 21:905–940.

Brown, R., 1987, From groups to groupoids: a brief survey, Bulletin of the London Mathematical Society, 19:113–134.

Cover, T., J. Thomas, 2006, Elements of Information Theory, Second Edition, Wiley, New York.

de Groot, S., P. Mazur, 1984, Non-Equilibrium Thermodynamics, Dover, New York.

Derman, E., M. Miller, D. Park, 2016, The Volatility Smile, Wiley, New York.

Diamond, D., A. Campbell, C. Park, J. Halonen, P. Zoladz, 2007, The Temporal Dynamics Model of Emotional Memory Processing Neural Plasticity, https://doi.org/10.1155/2007/60803.

Dolan, B., W. Janke, D. Johnston, M. Stathakopoulos, 2001, Thin Fisher zeros, Journal of Physics A 34:6211–6223.

Dretske, F., 1994, The explanatory role of information, Philosophical Transactions of the Royal Society A 349:59–70.
Dupy, T., 1977, A Genius for War: The German Army and General Staff 1807–1945, Macdonald and Janes, London.
El-Gohary, A., A. Alshamrani, A. Naif Al-Otaibi, 2013, The generalized Gompertz distribution, Applied Mathematical Modelling 37:13–24.
Feynman, R., 2000, Lectures on Computation, Westview Press, New York.
Fisher, M., 1965 Lectures in Theoretical Physics vol. 7, University of Colorado Press, Boulder.
Gerstner, L, 2003, Culture is Everything, Harvard Business Review, https://www.alumni.hbs.edu/stories/Pages/story-bulletin.aspx?num=2039.
Harush, U., B. Barzel, 2017, Dynamic patterns of information flow in complex networks, Nature Communications 8:2181
Jackson, D., A. Kempf, A. Morales, 2017, A robust generalization of the Legendre transform for QFT, Journal of Physics A 50:225201.
Jin, H., Z. Hu, X. Zhou, 2008, A convex stochastic optimization problem arising from portfolio selection, Mathematical Finance 18:171–183.
Khinchin, A., 1957, The Mathematical Foundations of Information Theory, Dover, New York.
Laidler, K., 1987, Chemical Kinetics, Third Edition, Harper and Row, New York.
Landau, L., E. Lifshitz, 2007, Statistical Physics, 3rd Ed. Part 1, Elsevier, New York.
Lerner, M., 2024, Workers crave autonomy and flexibility. Here are ways to achieve that balance, American Psychological Association, https://www.apa.org/topics/healthy-workplaces/work-life-harmony.
Maturana, H., F. Varela, 1980, Autopoiesis and Cognition: The Realization of the Living, Reidel, Boston.
Matzenbacher, B., 2018, The US Army and Mission Command: Philosophy versus Practice, Military Review, March-April, 61–71.
Mezo I., G. Keady, 2015, Some physical applications of generalized Lambert functions, arXiv:1505.01555v2 [math.CA] 22 Jun 2015.
Mintzberg, H., 1973, The Nature of Managerial Work, Harper and Row, New York.
Mintzberg, H., 1989, The Structure of Organizations, pp. 322–352, Readings in Strategic Management, Macmillan, UK.
Nair, G., F. Fagnani, S. Zampieri, R. Evans, 2007, Feedback control under data rate constraints: an overview, Proceedings of the IEEE 95, 108138.
Nisbett, R., K. Peng, C. Incheol, A. Norenzayan, 2001, Culture and systems of thought: holistic vs. analytic cognition, Psychological Review 108:291–310.
Osinga, F., 2007, Science, strategy and war: the strategic theory of John Boyd, Routledge, London.
Pettini, M., 2007, Geometry and Topology in Hamiltonian Dynamics and Statistical Mechanics, Springer, New York.
Porter, M., 1980, Competitive Strategy, Free Press, New York.
Porter, M., 2008, The Five Competitive Forces that Shape Strategy, Harvard Business Review 86:78–93.
Reina, A., T. Bose, V. Trianni, J. Marshall, 2018, Psychophysical laws and the superorganism, Scientific Reports, 8:4387, https://doi.org/10.1038/s41598-018-22616-y
Robinson, S., 1993, Shadow prices for measures of effectiveness II: General model, Operations Research 41:536–548.
Ruelle, D., 1964, Cluster property of the correlation functions of classical gases, Reviews of Modern Physics, April, 580–584.
Storr, J., 2007, A command philosophy for the information age: The continuing relevance of mission command, Defense Studies 3:119–129, https://doi.org/10.1080/14702430308405081.
Taleb, N. N. (ed.), 2020, Statistical Consequences of Fat Tails: Real world preasymptotics, epistemology, and applications, STEM Academic Press.
Wallace, R., 2021. How AI founders on adversarial landscapes of fog and friction, Journal of Defense Modeling and Simulation https://doi.org/10.1177/1548512920962227.

Wallace, R., 2022, Consciousness, Cognition, and Crosstalk: The evolutionary exaptation of nonergodic groupoid symmetry-breaking, Springer, New York.
Wallace, R., 2023, Essays on the Extended Evolutionary Synthesis: Formalizations and expansions, Sprnger, New York.
Wallace, R., G. Fricchione, 2024, Stress-induced failure of embodied cognition: a general model, BioSystems, https://www.sciencedirect.com/science/article/abs/pii/S0303264724000789.
Watts, D., S. Strogatz, 1998, Collective dynamics of small world networks, Nature 393:440–442.
Weinstein, A., 1996, Groupoids: unifying internal and external symmetry, Notices of the American Mathematical Association, 43:744–752.
Yi, S., P.W. Nelson and A.G. Ulsoy, 2010, Time-Delay Systems: Analysis and Control Using the Lambert W Function, World Scientific, New Jersey.
Zilberstein, S., 1996, Using anytime algorithms in intelligent systems, AI Magazine, 17:73–83.

Chapter 10
The Psychopathologies of Cultural Artifacts

Culture is as much a part of human biology as the enamel on our teeth.

—Boyd and Richerson (2005).

There is, as usual, an embedding context to what we have done here, although discussion has focused so far only on the singular failure inherent to a Western 'Tech Bro' culture-bound conception of machine cognition/artificial intelligence that views cognitive processes in isolation from their own embedding contexts. Given the ubiquity of Western Tech Bro ideologies across most current intellectual discipline and praxis, let us take something of our own medicine.

What, then, are these so-called 'cultural artifacts'?

The first, of course, are individual human minds. This has long been a central matter for cultural anthropology. As Gendron et al. (2020) put it,

> Cultural transmission prepares the developing brain and body to meet recurrent demands within a particular cultural context, thereby supporting the development of an internal model that is sufficiently tuned to specific environments. In this way, a human brain becomes wired to run a model of the world that will control the body in an efficient, predictive manner…[In essence,] a human brain becomes a cultural artifact…
>
> …Neuroscience advances in our understanding of the structure and organization of the human brain (including evolutionary (dis)continuities) strongly suggests that human culture wires the brain with the necessary flexibility and complexity to contend with the expansiveness of the human ecological niche.

Barrett et al. (2012) comment on similar matters:

> Our minds are constantly being transformed over the course of our individual development, and also over historical time, by the inclusion of psychological tools, such as language, and material culture into our daily lives. Viewed as a process, it becomes clear that our minds have not just been shaped by evolution, but also have been formed by those sociocultural processes that occur during development, processes that have, and will, vary in both time and space…Human cognitive evolution is, then an ongoing and contingent process.

R. Wallace, *Hallucination and Panic in Autonomous Systems*, Studies in Computational Intelligence 1217, https://doi.org/10.1007/978-3-031-95207-4_10

Wallace (2015), focusing on the failure of 'Western' perspectives, writes

> The stabilization of human cognition via feedback from embedding social and cultural contexts is a dynamic process deeply intertwined with it, constituting, in a sense, the riverbanks directing the flow of a stream of generalized consciousness at different scales: Cultural norms and social interaction are synergistic with individual and group cognition and their disorders. A canonical failure mode in atomistic cultures is found to be a 'ground state' collapse well represented by atomistic models of economic interaction that are increasingly characterized as divorced from reality by heterodox economists. That is, high rates of psychopathic and antisocial personality disorder and obsessive compulsive disorder emerge as culture-bound syndromes particular to Western or Westernizing societies, or to those undergoing social disintegration.

Lewis-Fernandez (2016) encapsulates this as follows,

> Culture shapes every aspect of patient care in psychiatry, influencing when, where, how, and to whom patients narrate their experiences of illness and distress…the patterning of symptoms…and the models clinicians use to interpret and understand symptoms in terms of psychiatric diagnoses…Culture also shapes patients' perceptions of care, including what types of treatment are acceptable and for how long…Even when patients and clinicians share similar cultural, ethnic or linguistic backgrounds, culture impacts care through other influences on identity, such as those due to gender, age, class, race, occupation, sexual orientation, and religion…Cultural contexts and expectations frame the clinical encounter for every patient, not only underserved minority groups, and cultural formulation therefore is an essential component of any comprehensive psychiatric assessment…
>
> The influence of culture on psychiatric presentations can be profound.

Generalizing, Wallace (2018) expands the argument:

> Cognition in living entities—and their social groupings or institutional artifacts—is necessarily as complicated as their embedding environments, which, for humans, includes a particularly rich cultural milieu. The asymptotic limit theorems of information and control theories permit construction of a new class of empirical 'regression-like' statistical models for cognitive developmental processes, their dynamics, and modes of dysfunction. Such models may, as have their simpler analogs, prove useful in the study and remediation of cognitive failure at and across the scales and levels of organization that constitute and drive the phenomena of life. These new models particularly focus on the roles of sociocultural environment and stress, in a large sense, as both trigger for the failure of the regulation of biocognition and as 'riverbanks' determining the channels of pathology, with implications across life-course developmental trajectories. [It is possible to] examine the effects of an embedding cultural milieu and its socioeconomic implementations using the 'lenses' of metabolic optimization, control system theory, and an extension of symmetry-breaking appropriate to information systems. A central implication is that most, if not all, human developmental disorders are fundamentally culture-bound syndromes. This has deep implications for both individual treatment and public health policy.

The next level, as discussed briefly in the earlier chapters, involves the cognitive institutions that have paralleled our rise over the last million years as the most successful predatory species on the planet—forgetting, for the moment, pandemic disease.

Wallace (2020, p.151), Sisyphus-like, continues the contextual analysis at the institutional level:

> Conflict between cognitive institutions on a Clausewitz landscape dominated by friction and the fog-of-war enters the same realm of difficulty as confronts and afflicts the study of consciousness in higher animals…Cognitive submodules of institutions and higher animals become linked, at different scales and levels of organization, in tunable, dynamically shifting coalitions in response to signals arising from both internal crosstalk and external 'sensory' sources. The formal description of consciousness remains highly contentious and appears to require mathematical tools familiar to the far outlands of string theory. This is no exaggeration: the art and science of war is, perhaps, even more confounding than our most arcane scientific edifices.

Gray (2006), in a military context, put the matter of Western institutional cultural siloing as follows:

> The serious trouble begins when Americans have to interact with alien societies in unfamiliar terrain…A certainty of material superiority can breed overconfidence and minimize incentives to outthink the enemy…A practical consequence of the messianic, crusading dimension to American strategic behavior, is a self-righteousness that is not friendly, or even receptive, to unwelcome cultural and political facts…The country's culture simply did not register the unwanted Vietnam experience. Cultures, including strategic cultures, are capable of ignoring what they wish to ignore.

Those, like the author, embedded in an American medical context, will, perhaps, recognize Gray's characterization.

Individual human minds, human institutions—from families through the United Nations—and our devices of machine cognition, are all cultural artifacts. As Sect. 1.2 above laments, Western cultural constraints that emphasize salience at the expense of context—mental disorders are 'in the brain' and 'in the genes', economics is about maximizing corporate profit, driverless cars are atoms in a traffic stream—seriously hinder our abilities to meet real-world challenges under Clauzewitzian fog, friction and deadly adversarial intent. That cultural straitjacket already limits our abilities to respond to an acute and growing range of tangled, self-reinforcing selection pressures, from human-induced climate change and environmental degradation to human-induced pandemic disease.

We are, then, increasingly confronted by what are effectively culture-bound syndromes of cognitive failure at and across the various scales and levels of organization of the human enterprise. These pathological dynamics entrain our many 'cockpit' partnerships, encompassing individual human minds, small groups, institutions, and newly-emerging machine entities that, to reiterate, are each and all autonomous cultural artifacts. Those affected then become hypersusceptible to 'Maskirovka' interventions by adversaries, true-believer proselytizers, or other con men (e.g., Wallace 2022).

It is no great surprise that Western fleets of Venture Capital hunting the Great White Whale of Artificial General Intelligence are foundering on these same shoals.

"But you can eat as much as you want from the tree of bullshit."

Licensed from The New Yorker, 2024

References

Barrett, L. S.P. Henzi, D. Lusseau, 2012, Taking sociality seriously: the structure of multi-dimensional social networks as a source of information for individuals, Philosophical Transactions of the Royal Society B, 367:2108–2118, https://doi.org/10.1098/rstb.2012.0113.

Boyd, R., P. Richerson, 2005, The origin and evolution of cultures, Oxford University Press, New York.

Gendron, M., B. Mesquita, L. Barrett, 2020, The Brain as a Cultural Artifact: Concepts, Actions, and Experiences within the Human Affective Niche, Chapter 6 in Kirmayer et al. (eds.), Culture, Mind, and Brain, Cambridge University Press, New York.

Gray, C.S., 2006, Out of the wilderness: Prime time for strategic culture. Prepared for the Defense Threat Reduction Agency Advanced Systems and Concepts Office, Contract Nos. DTRA01-03-D-0017, SP0600-04-C-5982.

Lewis-Fernandez, R., 2016, Culture and psychiatric classification, World Cultural Psychiatry Review 11:1–11.

Wallace, R., 2015, Closed-system 'economic' models for psychiatric disorders: Western atomism and its culture-bound syndromes, Cognitive Processing 16:279–290.

Wallace, R., 2018, Culture and the trajectories of developmental pathology: Insights from control and information theories, Acta Biotheoretica, https://doi.org/10.1007/s10441-018-9320-4.

Wallace, R., 2020, Cognitive Dynamics on Clausewitz Landscapes: The control and directed evolution of organized conflict, Springer, New York.

Wallace, R., 2022, Deception and Delay in Organized Conflict: Essays on the mathematical theory of Maskirovka, Springer, New York.